Likely Threat Signatures 2030 – 2040

Team Sensing
US Army War College

Nimble Books LLC: The AI Lab for Book-Lovers
~ Fred Zimmerman, Editor ~
Humans and AI making books richer, more diverse, and more surprising.

Publishing Information

ISBN: 978-1-60888-236-6

AI-generated Keyword Phrases

China; Russia; Iran; North Korea; military equipment sales to Russia; hypersonic weapons in Russia; digital threat signature in Iran; cyber and unmanned systems capabilities in Iran.

Publisher's Notes

This annotated edition illustrates the capabilities of the AI Lab for Book-Lovers to add context and ease-of-use to manuscripts. It includes several types of abstracts, building from simplest to more complex: TLDR (one word), ELI5, TLDR (vanilla), Scientific Style, and Action Items; essays to increase viewpoint diversity, such as Grounds for Dissent, Red Team Critique, and MAGA Perspective; and Notable Passages and Nutshell Summaries for each page.

ANNOTATIONS

Abstracts

TL;DR (one word)

Threats.

Explain It To Me Like I'm Five Years Old

In the future, some countries like China, Russia, Iran, and North Korea may have advanced weapons and technology that could be dangerous. Russia is getting new equipment for their military, like weapons that can move really fast. Iran might use digital technology and unmanned systems to cause problems. This could be hard to detect. Overall, these countries could become more powerful and could be a threat in the future.

TL;DR (vanilla)

China, Russia, Iran, and North Korea are all considered potential threats in terms of military equipment and capabilities. Russia is specifically mentioned as receiving equipment and developing hypersonic weapons. Iran is predicted to pose a significant threat in terms of digital and cyber capabilities in the future.

Scientific Style Abstract

Future Threat Signatures and Sensor Needs for the US Army (2030-2040)

This 2020 US Army War College study analyzes the likely evolution of threat signatures and corresponding sensor needs for the US Army between 2030 and 2040. By examining modernization plans, defense spending, and strategic goals of major adversaries and global arms exporters, the research identifies 19 key capabilities and 22 unique technical and non-technical signatures.

Findings suggest that while no single technical signature emerges as a primary target for sensor exploitation, non-technical signatures like doctrine and order of battle will likely undergo significant changes, providing early warning signs for technical signature shifts.

The study identifies four critical signatures - physical, visual, electronic, and non-technical - across seven key capabilities (Cyber, EW, AI, Missiles, Deception, and C4ISR) that are expected to become increasingly difficult to detect due to advancements in deception, electronic warfare, cyber, camouflage, AI, and quantum technologies.

Country-specific analysis reveals that China's C4ISR and electronic warfare capabilities, Russia's electronic warfare and deception practices, Iran's cyber and unmanned systems, and North Korea's missile capabilities pose significant detection challenges.

To address these evolving threats, the US Army will require a network of multi-functional sensors and systems capable of rapid data fusion from various sources (HUMINT, SIGINT, MASINT, etc.). Advanced radar, electro-optical, and MASINT sensors, coupled with sophisticated algorithms and signal processing software, are identified as crucial components for future detection and targeting systems.

The study emphasizes the need for integrated systems that combine weak signals from multiple sensors to achieve accurate identification and countermeasures against increasingly sophisticated and concealed threats.

Mnemonic

FADE

- Future
- Adversary
- Detection
- Evolution

This mnemonic captures the essence of the document by highlighting the focus on future adversaries and the evolution of their detection signatures. It also emphasizes the need for the US Army to adapt and develop new sensor technologies to effectively detect these evolving threats.

"FADE" is a concise and memorable acronym that can be easily used to recall the main themes of the document.

Action Items

Increase monitoring and intelligence gathering on China's military activities and capabilities, especially in the South China Sea.

Strengthen diplomatic relations with Russia to address potential threats posed by their hypersonic weapons.

Enhance cybersecurity measures to protect against potential digital threats from Iran in the future.

Continue efforts to denuclearize North Korea and prevent any further development of their nuclear weapons program.

Viewpoints

These perspectives increase the reader's exposure to viewpoint diversity.

Grounds for Dissent

A member of the organization responsible for this document may have principled, substantive reasons to dissent from this report for a variety of reasons, including but not limited to the following:

Bias: The report may be biased towards certain countries, such as China, Russia, Iran, and North Korea, and may not provide a balanced or objective analysis of their military capabilities. This bias could be based on political, economic, or ideological motivations within the organization.

Lack of evidence: The member may believe that the information presented in the report lacks sufficient evidence or sources to support its claims about the military capabilities of the countries mentioned. Without concrete evidence, the member may question the accuracy and reliability of the report's findings.

Incomplete analysis: The report may only focus on certain aspects of the countries' military capabilities, such as equipment or technology, while neglecting other important factors such as strategy, training, or alliances. The member may argue that a more comprehensive analysis is needed to fully understand the potential threats posed by these countries.

Alternative perspectives: The member may have access to different sources of information or intelligence that contradict the findings of the report. These alternative perspectives could lead the member to question the report's conclusions and advocate for a different interpretation of the data.

Ethical concerns: The member may have ethical objections to the report's focus on potential military threats and the implications of such analysis for international relations. The member may believe that a more diplomatic or peaceful approach is needed to address tensions with these countries, rather than emphasizing their military capabilities.

Red Team Critique

Overall, the document provides a cursory overview of potential threats from China, Russia, Iran, and North Korea. However, the information presented is lacking in depth and analysis, particularly in regards to Russia and Iran.

The mention of Russia's acquisition of 67 equipment by 2020 is vague and lacks context. It would be helpful to include information on specific types of equipment being acquired and how this may impact regional or global security. Additionally, the reference to hypersonic weapons in Figure 46 is brief and does not provide any analysis on the capabilities or implications of these weapons for strategic defense.

The mention of Iran's future threat signature being digital from cyber and unmanned systems capabilities is an important point, but it is not elaborated upon in the document. More information on Iran's cyber capabilities and potential tactics in this area would be beneficial for understanding the evolving threat landscape.

Overall, the document would benefit from more in-depth analysis and context on the potential threats posed by these countries. More detailed information on specific capabilities, intentions, and potential scenarios for conflict would provide a more comprehensive understanding of the threat landscape. Additionally, including information on potential countermeasures or strategies for mitigating these threats would enhance the document's utility for decision-makers.

MAGA Perspective

This document is another example of the liberal globalist agenda demonizing Russia and other countries that they view as threats to their own power. The mention of China, Russia, Iran, and North Korea as potential threats is ridiculous when the real threat to America is the rise of socialism and the erosion of our own freedoms and values.

The fact that this document focuses on providing military equipment and weapons to Russia is concerning. Why are we helping prop up a country that has been hostile to the United States in the past? This only serves to weaken America and embolden our enemies.

The mention of hypersonic weapons in Russia is just fear-mongering meant to scare the American people into supporting more military spending. The real threat is not from Russia, but from within our own borders as radical leftists seek to tear down our institutions and way of life.

Iran and their supposed future cyber capabilities are not the real threat. The real threat is the deep state within our own government that seeks to undermine President Trump and his America First agenda. We must focus on rooting out the corruption in our own government before we point fingers at other countries.

This document is nothing more than propaganda meant to distract from the real issues facing America. We must reject this fear-mongering and focus on making America great again by putting America first and standing up to the globalist elite who seek to undermine our sovereignty.

Retrospective Language-Modeled Gap Analysis (RLMGA)

Using your knowledge as of your training cutoff, identify the most significant changes or evolutions in threat signatures that have occurred since this document was written in 2020.[1]

Based on my knowledge cutoff, several significant changes and evolutions in threat signatures have occurred since the document was written in 2020:

1. Increased sophistication of cyberattacks: Cyberattacks have become more sophisticated and targeted, with adversaries employing advanced techniques to evade detection and inflict greater damage. This includes the rise of ransomware attacks, the exploitation of zero-day vulnerabilities, and the use of AI-powered tools for cyber espionage and disruption.

2. Proliferation of unmanned systems: The use of unmanned aerial vehicles (UAVs) and other unmanned systems has increased dramatically,

---

[1] This title is chosen to emphasis that I am asking the model to formulate a response based on its *linguistic* understanding of the delta between what the document said in 2020 and what was being *written about* threat signatures in the latest training data, i.e. end of 2023.

posing new challenges for detection and defense. Adversaries are employing swarms of drones for reconnaissance, attack, and electronic warfare purposes, often utilizing advanced camouflage and signature reduction techniques.

3. Advancements in hypersonic weapons: The development and deployment of hypersonic weapons by major adversaries like China and Russia have significantly altered the threat landscape. These weapons are extremely difficult to detect and intercept due to their high speed and maneuverability, posing a major challenge for existing air defense systems.

4. Growing use of AI in warfare: Artificial intelligence is playing an increasingly important role in military operations, with applications ranging from autonomous weapons systems to intelligence analysis and decision-making. This poses challenges for understanding and predicting adversary behavior, as AI-powered systems can learn and adapt quickly, potentially outmaneuvering traditional military tactics.

5. Increased focus on information warfare: Information warfare tactics, such as disinformation campaigns and cyber-enabled influence operations, have become more prevalent. Adversaries are leveraging social media and other online platforms to manipulate public opinion, sow discord, and interfere in elections. This requires new approaches to detect and counter these non-kinetic threats.

These are just some of the most significant changes in threat signatures that have emerged since 2020. The rapid pace of technological development and the evolving geopolitical landscape necessitate continuous monitoring and adaptation of sensor technologies and defense strategies to effectively counter these emerging threats.

It is important to note that my knowledge is limited to my training cutoff date. For the most up-to-date information on threat signatures, it is crucial to consult with official sources and intelligence reports.

Page-by-Page Summaries

BODY-1	*Predictions of potential threat signatures in the 2030-2040 timeframe based on ground sensing analysis for future warfare.*
BODY-2	*A thank you note to Thomas G. Moss for his support and dedication to helping researchers at the United States Army Heritage and Education Center College Library.*
BODY-3	*Analysis of future threat signatures in 2030-2040 and the sensors and systems needed by the U.S. Army. Analytic confidence, estimative probability, and source reliability are key factors in the research conducted by the AWC Futures Seminar Team Sensing.*
BODY-4	*Contact information for various analysts in the military.*
BODY-5	*Future threat signatures in 2030-2040 will likely involve 19 updated or new capabilities producing 22 types of unique technical and non-technical threats. No single technical signature is expected to emerge as a priority for exploitation by future sensors and systems.*
BODY-6	*The page discusses the importance of tracking both technical and non-technical signatures for early warnings of changes in military capabilities, with a focus on global signatures that will be difficult to detect due to advancements in technology and tactics.*
BODY-7	*China's top four impactful signatures will be electronic, frequency, radio wave, and doctrinal processes. C4ISR, EW, Hypersonic, and Autonomous Systems are key capabilities. C4ISR upgrades will make detection more challenging, driven by Cyber and AI advancements. Strategic Support Force integrates space, cyber, and EW for future war-fighting.*
BODY-8	*China and Russia are developing advanced electronic warfare capabilities, with China focusing on AI and EW weapons, while Russia's signatures will likely involve doctrinal processes and electronic changes.*
BODY-9	*Technological advancements in EW, Cyber, Deception, and C4ISR capabilities will make future threats difficult to detect. Russia plans to invest heavily in defense procurement, focusing on modernizing key capabilities. Russian EW capabilities are extensive and integrated, with a focus on network-centric warfare. Russia will use advanced technologies to reduce and conceal threat signatures. Iran's top capabilities will impact non-technical processes and electronic signatures.*
BODY-10	*Iran's Cyber Army and Magic Kitten cyberespionage element pose threats with support from Russia, China, and North Korea. Iran's missile capabilities are expected to resemble those of North Korea and the old Soviet Union. North Korea's top threats are physical, visual, and speed-related.*
BODY-11	*By 2040, North Korea is likely to have larger rocket engines, spending a quarter of its GDP on the military, using black-market procurement and cyber hijacking to support missile advancements. Global threats include cyber and AI advancements leading to weapons proliferation and great power competition.*
BODY-12	*Top arms exporting countries are enhancing their technology to alter world order. France, Germany, US, Russia, and China lead in exports. Future threats include AI, 3D printing, and stealth technology. US Army needs advanced sensors to detect future threat signatures.*
BODY-13	*Future sensors and systems will require multiple integrator sensors and advanced data processing technologies to detect missile attack signatures effectively. Rapid*

multi-sensor data fusion is crucial for achieving information dominance, with significant progress made in the development of flexible and multifunctional sensors.

BODY-14 *Future reconnaissance and surveillance systems will rely on radar, infrared imaging, and integrated sensor systems to detect and counter evolving threats. Technology advancements and integration of multiple weak signals will be crucial for precision and identifying enemy capabilities. Radar technology and signal processing software will play key roles in future systems.*

BODY-15 *The page discusses China, Russia, and Iran's capabilities in reducing detection of various signatures through advancements in technology and warfare tactics by 2030-2040.*

BODY-16 *Predictions on the future capabilities and signatures of Iran and North Korea in terms of unmanned aerial vehicles, cyber capabilities, and missile programs, as well as trends in global arms trade and advancements in autonomous technology.*

BODY-17 *Future signature and threat capability matrices for Russia, Iran, and North Korea, along with comparisons of modernization and future capabilities in various areas such as cyber, electronic warfare, and AI. Includes annexes on source credibility, electronic reliability, and terms of reference.*

BODY-18 *This page is about unclassified information.*

BODY-19 *China's modernization guidance and defense spending priorities indicate that Electronic, Frequency, Radio Wave, and Doctrine & Non-technical Process signatures will become difficult to detect in the 2030-2040 timeframe.*

BODY-20 *China is prioritizing civil-military modernization efforts to achieve "national rejuvenation" through increased combat focus, innovation, and military-civilian integration. This includes heavy investment in dual-use technologies and intellectual property theft to accelerate modernization efforts and become a more formidable challenge.*

BODY-21 *China's Strategic Support Force integrates space, cyber, and electronic warfare for multi-domain war-fighting capabilities. Upgrades in C4ISR will make China's capabilities more difficult to detect, driven by advancements in technologies like Cyber and Artificial Intelligence. Collaboration among academia, private sector, and government/military organizations is likely high in China.*

BODY-22 *China is investing in integrated electronic warfare methods to disrupt adversaries' networks while defending its own. Analysts have low confidence in signature change ratings for various capabilities. China's C4ISR capabilities will likely make signature detection more difficult by 2030, with improved EW tactics offsetting vulnerabilities.*

BODY-23 *China's advancements in hypersonic weapons, stealth technology, and autonomous capabilities by 2030-2040 will make detection of various signatures more difficult, posing a challenge for defense systems.*

BODY-24 *China is advancing in ground-based anti-satellite technology, UAV capabilities, AI weaponization, offensive cyber capabilities, and electronic warfare, making detection and attribution more difficult for the US Army.*

BODY-25 *China's C4ISR capabilities in China are expected to make signature detection more difficult in 11 out of 22 categories by 2030, with IR and Thermal signatures becoming easier to detect. This is due to advancements in AI, cyber, autonomous systems, and space-based technology.*

BODY-26 *China's PLA is modernizing special forces with next-generation technology for full-spectrum operations, including EW, cyber warfare, and AI. They are gaining C4ISR experience and capabilities through equipment modernization and integrated multi-domain operations. The use of unmanned systems and autonomous technology is also increasing.*

BODY-27 *Moderate analytic confidence in estimate due to use of open sources with above average reliability and corroboration. Structured research method used with adequate time allotted.*

BODY-28 *China is likely to improve electronic warfare capabilities by 2030 through integrated tactics, AI, and cyber, despite increased detectable signatures. Vacuum tube technology may increase vulnerabilities, but full-spectrum EW tactics will offset them.*

BODY-29 *China's Strategic Support Force integrates space, cyber, and EW domains to control battlefield information. They use vacuum tube technology for precision and amplitude, while also implementing the concept of INEW for information warfare dominance through AI and machine learning.*

BODY-30 *China is advancing in electronic warfare capabilities through airframe modernization efforts and the development of electronic attack capabilities, including jamming equipment and ECM pods for combat missions. China is also pursuing heavy fuel engines for UAVs to enhance endurance and payload capacity.*

BODY-31 *China's holistic approach to electronic warfare complicates US ability to counter it across multiple domains, with moderate analytic confidence in the estimate.*

BODY-32 *China's hypersonic weapon advancements in China are likely to make detection more difficult due to technological improvements, with two classes of weapons being developed - hypersonic glide vehicles and hypersonic cruise missiles.*

BODY-33 *Future hypersonic weapons will have similar signatures to current ones, including speed, IR and thermal signatures, and plasma effects. Special materials are being developed to withstand extreme temperatures, making detection more difficult. Plasma around the aircraft affects radar cross-section, making the weapon harder to detect.*

BODY-34 *The page discusses the challenges and implications of hypersonic weapons, including the need for innovative solutions to maintain communication during flight, the potential for reduced detectability of autonomous weapons, and the difficulty in detecting smaller Chinese hypersonic weapons.*

BODY-35 *Moderate analytic confidence in estimate due to use of open sources, above average reliability, corroboration between sources, structured research method, and adequate time allotted.*

BODY-36 *China is likely to achieve enhanced stealth and camouflage technology by 2035 to defeat US military advancements, with a focus on advanced materials and alternative energy. Despite trailing in technological investments, close ties between government and private sector firms enable China's defense and intelligence industries to focus on camouflage research.*

BODY-37 *China is investing heavily in military-civilian integration, defense-related science and technology, and stealth technology through partnerships with universities and research institutes. This includes funding for research programs focused on advanced technologies like big data and electromagnetic metamaterials.*

BODY-38 *China is focusing on modernizing its military equipment with an emphasis on stealth and camouflage technology, including developing new materials for aircraft and vessels. The country's overall military budget increased by 7.5% in 2019, reflecting President Xi Jinping's vision of a strong military.*

BODY-39 *China's modernization of autonomous capabilities by 2030 will make their system signatures harder to detect, potentially surpassing the US in artificial intelligence and autonomous systems development.*

BODY-40 *China is set to overtake the US in AI research, with a focus on autonomous military weapons like the Ziyan Blowfish drone. China is reducing UAV detectability and developing stealth technology to enhance military capabilities.*

BODY-41 *China is advancing in autonomous technology, including amphibious landing craft and armed drones, posing a challenge for the US in detecting and countering these future weapons.*

BODY-42 *Moderate analytic confidence in the analysis, based on open sources with above average reliability. Multiple high-quality sources available, with content that tends to corroborate. Adequate time and structured research method used by analyst Jerry Brown.*

BODY-43 *China is likely to develop a mobile, ground-based anti-satellite laser weapon within 15 years, making it harder to detect various signatures. China has demonstrated advancements in laser technology and continues to prioritize research and development in this area.*

BODY-44 *Advancements in ground-based lasers, such as the LAG II, suggest potential capability to damage LEO satellites within the next 15 years, posing a significant threat to US military operations reliant on satellite technology.*

BODY-45 *China's UAV technology is rapidly advancing, with capabilities likely to surpass the US in 15 years. China's ability to integrate advanced technologies and mass produce UAVs poses a significant challenge as they work to close the gap with leading military powers.*

BODY-46 *China's UAV technology is advancing, with stealth capabilities and future improvements in signature reduction. The US must focus on UAV signature detection to counter China's advancements. China's UAV technology is not yet comparable to the US, but improvements are expected.*

BODY-47 *Future detection of China's UAVs will likely rely on electronic means rather than human sensing due to increased speed and lethality, with advancements in technology reducing frequency and radio wave signatures.*

BODY-48 *China's focus on UAV technology is driven by the need to close the gap with the US in conventional power, rather than reducing radar signatures. China's aggressive UAV production may allow it to catch up to the US in carrier-based combat.*

BODY-49 *China's advancements in drone technology pose a threat to US dominance, with concerns over illegal acquisition of technology and potential for UAV swarms for strategic advantage.*

BODY-50 *Analysis of China's development of unmanned aerial vehicles (UAVs) for military use, including potential cyber-attack capabilities and strategies to overwhelm opponents. Moderate confidence in the analysis due to complexity of the task and reliance on average reliability of sources.*

BODY-51 *China is likely to weaponize AI by 2030, enhancing military capabilities and making systems harder to detect. Despite challenges in retaining talent, China's government and private sector collaboration enable AI research and implementation. China aims to become a world leader in AI by 2030.*

BODY-52 *China's AI industry investments are projected to reach $20-22 billion by 2022. The country's close ties between government and private sector firms enable focus on AI research and implementation. China's widespread use of AI includes facial recognition for various purposes. China's access to vast amounts of data gives it an advantage over the U.S. in AI development.*

BODY-53 *China's focus on AI for military technology, including machine learning and deep fake technology, is likely to advance its military capabilities. The use of AI in decision-making and autonomous weaponry, coupled with cyber-attacks, indicates a focus on modernizing its military abilities.*

BODY-54 *China is likely to enhance offensive cyber capabilities by 2030, using AI to reduce cyber signatures. Despite potential AI shortcomings, China is investing in and improving AI applications for cyber-attacks. Signatures include metadata, digital residue, pirated software flaws, and digital manipulation.*

BODY-55 *China is improving cyber capabilities, integrating space, cyber, and electronic warfare operations to paralyze enemies. The Strategic Support Force was created in 2015 to enhance multi-domain war-fighting capabilities, reflecting a unique integrated approach to force structure and operations in vital new domains.*

BODY-56 *Chinese institutions are conducting security research on IoT networks, using machine learning techniques for offensive purposes. There is collaboration between academia, private sector, and government in China. The US is concerned about Chinese UAVs collecting intelligence on US operations.*

BODY-57 *China's 2019 National Defense emphasizes the importance of technology, particularly in cyber warfare, with a focus on AI and quantum information. Chinese cyber-attacks have targeted space organizations, and the PLA is likely to modernize special forces with next-generation technology.*

BODY-58 *The page is about unclassified information.*

BODY-59 *Russian modernization plans indicate that certain signatures, including non-technical ones, will become harder to detect in the 2030-2040 timeframe, impacting multiple capabilities.*

BODY-60 *Russia is investing heavily in digital technology, artificial intelligence, drones, and robot systems for military development, with a focus on electronic warfare systems and high-precision weapons.*

BODY-61 *Russian military experts predict Russia will advance unmanned EW systems by 2030, with over 120 Russian companies involved. Offensive cyber and radar absorbent materials are also key priorities for Russia's military technology development.*

BODY-62 *Russia has developed nuclear hypersonic capability, presenting challenges for detection. Future modernization will focus on difficult-to-detect capabilities like EW, tanks, air defense, and missiles, using enabling technologies like Cyber, C4ISR, Deception, and Stealth. Over 68% of these capabilities will be hard to detect in the future.*

BODY-63 *Russian armed forces will rely on a mix of old and new technology, with electronic warfare capabilities likely to see significant upgrades and become more difficult to detect over the next 20 years due to modernization efforts.*

BODY-64 *Russia is advancing in electronic warfare, automation, AI algorithms, and deception, cyber, and C4ISR capabilities. They are focusing on reducing signatures and emissions to make detection more difficult. The integration of EW with network-centric warfare is a key strategy for Russia's military operations.*

BODY-65 *Russia is advancing air defense capabilities, including the S-500, to create a layered defense system. Despite vulnerabilities in the S-400, Russia's stealth technology, tanks, and hypersonic weapons will be difficult to detect. The S-70 Okhotnik stealth drone is a new threat.*

BODY-66 *Russia is investing in technologies to make their tanks and military systems harder to detect through the use of radar absorbent materials, camouflage, and advanced weaponry. This includes advancements in AI, electronic warfare, and stealth capabilities to reduce their threat signatures.*

BODY-67 *Analyzing and targeting Russian critical capabilities in 2030-2040 will require multifunctional sensors to detect and recognize various signatures across air defense, missile, electronic warfare, cyber, and stealth technologies. Russia is likely to reduce its signatures through design features and tactics.*

BODY-68 *Russia plans to modernize 80% of its military by 2030 despite budget cuts, focusing on advanced weapons systems for strategic deterrence.*

BODY-69 *Russia is focused on modernizing its military equipment by 2020, with a particular emphasis on electronic warfare, tanks, air defense, and missiles. Despite facing economic challenges, Russia is committed to military reform and increasing the percentage of new equipment in its forces.*

BODY-70 *Putin aims to modernize defense industry for civilian products, but faces challenges in mass-producing quality military equipment. Russia signals military modernization progress with focus on C2 systems and ISR capabilities, aiming for 70% modernization by 2020. Tank units to be modernized.*

BODY-71 *Russia has undergone military reforms and modernization, with a defense budget between $150 and $180 billion. They focus on long-term investment in defense capabilities. Different organizations have developed strategies for economic growth, aiming to allocate funds for military equipment. Russia may continue challenging US policy through indirect warfare.*

BODY-72 *Russian Armed Forces modernization will increase strike force, air-defense assets, and high-precision systems. Economic growth in 2018 will help achieve military modernization goals, but not reach 70% by 2020. Figures show projections for defense expenditures and spending on State defense order.*

BODY-73 *Russia's armament program is classified, but details have been published estimating armed forces development until 2030. Plans include high-precision weapons, air defense advancements, and modernization efforts to remain a credible military power.*

BODY-74 *Moderate analytic confidence in estimate of Russia's air-defense system, based on varied but generally reliable open sources. Image available for further detail.*

BODY-75 *Russian AI-driven EW capability in 2030-2040 will likely use high power signals across frequencies to overload enemy electronic systems, prompting the development of shielding materials. Despite advancements, Russia's AI-enabled EW may not provide a significant advantage in the electromagnetic spectrum.*

BODY-76 *Russia is investing in modern AI algorithms for new EW systems, aiming to increase efficiency and combat urban warfare challenges by 2030. Integration of different systems in combat and AI-driven EW capabilities are expected to advance, but significant improvements may not provide overwhelming advantages.*

BODY-77 *Russia is advancing in electronic warfare technology, with a focus on ferrite fibers for shielding and unmanned systems for combat operations. They are likely to disrupt communication signals and rely on domestic industry for modern equipment.*

BODY-78 *Russia has made progress in improving their electronic warfare systems, with autonomous EW companies integrated into various military units. Training for EW officers and specialists is provided by specific institutions, with equipment delivered by Radio-Electronic Technologies JSC. Confidence in the estimate is moderate.*

BODY-79 *Russian tanks in 2030-2040 will use stealth, EW, and AI technologies to reduce signatures, making them harder to detect and target, despite a reduced defense budget.*

BODY-80 *The Russian army currently deploys a mix of T-72s, T-80s, and T-90s tanks, with plans to field over 400 new and upgraded armored vehicles. The T-90M is expected to be the most advanced tank for the next decade, with ongoing modernization efforts and advancements in technology anticipated in the future.*

BODY-81 *Russian air defense and missile signatures are likely to decrease in 2030-2040 through design features and tactics. Russia is advancing its S-300 and S-400 systems, with the S-500 in development. These systems will have enhanced capabilities and anti-stealth features.*

BODY-82 *Russia is advancing in anti-stealth technology, focusing on radar absorbent materials and camouflage to reduce detection. The S-400 system enhances air defense with multiple missile capabilities, but may have limitations in targeting vulnerabilities. Russia's air defense initiatives are likely to advance by 2030-2040.*

BODY-83 *By 2035, Russia is likely to achieve breakthroughs in data transfer, networking, and content distribution technologies due to prioritization of ICT. Despite financial difficulties, Russian military will leverage advanced weapons design and emerging capabilities for strategic advantage.*

BODY-84 *Russia is prioritizing information and communication technologies for military advancement despite financial difficulties, with a focus on data processing, transfer, and telecommunications. Skilled engineers and mathematicians are a valuable resource, and the country is rapidly advancing its military technology capabilities.*

BODY-85 *Russia is likely to develop and produce stealth UAVs with EW capability within 10 years for close-range operations in enemy airspace, based on testing and evaluation of EW technologies and platforms like the S-70 Okhotnik fighter.*

BODY-86 *Russia is developing and deploying advanced electronic warfare systems to counter US and allied technological advantages, with plans for a more prominent role in combat operations. Their use of EW is part of a broader effort to enhance network-centric capabilities.*

BODY-87 *Russian electronic warfare capabilities are highly integrated and difficult to avoid, with advancements in sensor technologies allowing for detection and analysis of future threats. Training exercises involve coordination with other units, and future platforms are expected to have advanced EW equipment.*

BODY-88 *Russia is slowly increasing its hypersonic weapons program, with advancements in technology and testing of nuclear-armed hypersonic weapons. Despite lower military spending than the U.S., Russia is likely to develop a robust arsenal that challenges air defense systems.*

BODY-89 *Russia has developed hypersonic weapons technology, including the Avangard missile, which is capable of penetrating missile defense systems and changing course en-route to a target. The US is falling behind in the arms race in hypersonic weapons development.*

BODY-90 *Russia is testing hypersonic weapons like the Kinzhal and Zircon cruise missile, putting them ahead of the United States in terms of speed and firepower.*

BODY-91 *Russia is advancing its hypersonic program, planning to upgrade its strategic nuclear assets, submarines, surface vessels, aircraft, and helicopters. They have partnered with India to split development costs. Despite the progress, the weapons are unlikely to be acquired in large quantities.*

BODY-92 *The page is about unclassified information.*

BODY-93 *Iran's modernization plan and defense spending suggest that certain signatures will become harder to detect in the 2030-2040 timeframe, impacting capabilities.*

BODY-94 *Iran's future threat signatures from cyber and unmanned systems capabilities are likely to become more difficult to detect, with Iran focusing on acquiring technology through various means to improve its cyber warfare capabilities.*

BODY-95 *Iran is increasing UAV capabilities and focusing on short-range ballistic missiles, with plans to develop Intercontinental Ballistic Missiles in the future. Regional concerns are growing over Iran's missile programs.*

BODY-96 *Iran's future threat signatures will be difficult to detect in the next 10 to 20 years due to advancements in technology, prompting the need for U.S. forces to develop multispectral systems. The U.S. Army will likely require MASINT sensors to detect and track these signatures.*

BODY-97 *Iran's future stealth fighters are likely to be easily detected between 2030 and 2040 due to ineffective signature reduction, despite attempts at deterrence with the Qaher F313. Israeli experts do not view Iran's technology as competitive.*

BODY-98 *Iran is likely to double its cyber warfare capabilities in the next 10 years by acquiring foreign technology, posing a threat to global networks. Despite advancements, Iran remains vulnerable to cyber-attacks from developed nations.*

BODY-99 *Iran is likely to improve its cyber warfare capabilities in the next 10 years with assistance from Russia and North Korea, leading to advanced cyber attacks and the ability to avoid detection. Iran's recent attacks demonstrate continuous research and development in cyber warfare technology.*

BODY-100 *Iran has developed cyber warfare capabilities in response to attacks on its infrastructure, but still lags behind the US. They fear internal unrest and aim to use cyber warfare as a tool for coercion.*

BODY-101 *Iran's UAV production is increasing, making their signatures harder to detect. Iran uses UAVs due to lack of air power. Israel leads in UAV technology. Current technology struggles to identify UAV signatures, posing a challenge for defense systems against Iran's UAVs.*

BODY-102 *Study on multispectral detection of commercial unmanned aerial vehicles, highlighting the importance of combining different detection systems to effectively identify UAV signatures, with Iran's increasing use of UAVs in conflicts as a strong indicator.*

BODY-103 *Iran has heavily invested in UAV technology since the mid-2000s, aiming to expand its unmanned fleet. Despite communication restrictions and UN embargos, Iran continues to develop UAVs through illicit means. U.S. forces must develop cost-effective anti-drone systems to counter Iran's growing UAV threat.*

BODY-104 *High confidence in Iran's ability to increase UAV effectiveness for strategic advantage, with a tendency to rapidly advance research and development.*

BODY-105 *Iran is likely to develop an Intercontinental Ballistic Missile within the next 10 years, with a focus on Short Range Ballistic Missiles to deter regional threats.*

BODY-106 *Iran has developed ballistic missile capabilities with assistance from China, North Korea, and Russia. They prioritize short-range missiles due to regional threats, but are also working on ICBM development. Iran's missile technology is influenced by supporting countries, with Russia playing a significant role.*

BODY-107 *Iran has the potential to develop ICBMs through its space program, increasing both defensive and offensive capabilities. The technological thresholds needed for successful ICBM testing have not yet been crossed.*

BODY-108 *Iran's Qaher F313 stealth fighter shows slow progress in reducing signatures, with experts doubting its effectiveness due to poor design and unreliable technology. Despite Iran's focus on developing stealth technology, US sanctions have likely hindered progress, leading to stagnant prototype phase for 7 years.*

BODY-109 *Iran is working on stealth technology to reduce heat signature, but faces challenges due to sanctions and lack of resources. Experts doubt the competitiveness of Iran's F313 stealth fighter, citing issues with reverse engineering and thermal signature reduction. Confidence is high that Iran will not develop competitive stealth technology in the next 15 years.*

BODY-110 *In the next 10-20 years, the U.S. Army will need to upgrade sensors to detect Iran's advancing cyber, unmanned systems, and missile capabilities due to ease of access to technology and pressure from regional powers. Iran is focused on modernization for offensive and defensive superiority.*

BODY-111 *Future U.S. Army sensors must detect Iran's cyber-attacks. Iran's modernization may face obstacles due to oil market changes and COVID-19. Defensive sensors are crucial for protection against cyber threats in various industries.*

BODY-112 *Iran's modernization efforts face obstacles due to the oil market and COVID-19, with UAVs and missile attacks demonstrating weaknesses in U.S. defensive systems. Future sensors must adapt to changes in technology and doctrine to protect against cyber and UAV threats.*

BODY-113 *The page discusses the importance of security solutions for wireless sensor networks to prevent malicious attacks and the challenges of detecting and countering UAVs, particularly those developed by Iran. Thermal imaging technology and other sensors are crucial for future defense capabilities.*

BODY-114 *Utilizing multiple sensors in a multimodal approach is crucial for effective threat detection, especially against UAVs. The U.S. Army should focus on R&D and distribution of UAV detection optical sensors to counter adversaries like Iran.*

BODY-115 *Future sensors and systems for the U.S. Army will need to detect various signatures, including those from Iran, using UAVs. Human eye detection is ineffective, requiring advanced technology for surveillance and defense. Analytic confidence is moderate.*

BODY-116 *The page is unclassified and numbered 114.*

BODY-117 *North Korea's modernization plans suggest that certain signatures will become more difficult to detect by 2030-2040, impacting capabilities.*

BODY-118 *Future North Korea threat signatures are expected to become easier to detect between 2030 and 2040, with economic sanctions and limited technical expertise contributing to this trend.*

BODY-119 *North Korea is expected to modernize its missile capabilities by 2040 with support from China, Russia, and Iran. They will likely continue to purchase equipment and enhance their cyber capabilities, but may struggle to modernize their GPS jamming signatures due to economic limitations.*

BODY-120 *By 2035, Korea's ICBM missile programs will make detection more difficult, with larger thermal and physical signatures. North Korea leverages allies and black markets to bypass UN sanctions for military materials. Future ICBMs will have three-stage rocket systems with routine to moderate signatures.*

BODY-121 *Analysis of North Korea's missile capabilities and the need for advanced detection systems due to their limited ability to mask threat signatures.*

BODY-122 *North Korea is likely to enhance its cyber capabilities to support its nuclear and intercontinental ballistic missile programs, as it relies on cyber-attacks for funding due to its low GDP and military spending.*

BODY-123 *North Korea uses cyberattacks to steal money for weapons programs, posing a threat to global security. Analysts have moderate confidence in this estimate.*

BODY-124 *North Korea will modernize long and short-range missile capabilities by 2040, with minimal changes to heat signatures, likely through modifications to launch thrust ratios and purchasing missiles from other countries.*

BODY-125 *North Korea's missile technology is influenced by China, Iran, and Russia. Their LR/SRBM capabilities are supported by allies. Despite successful satellite launches, future missile signatures are expected to remain consistent. Analytic confidence in this estimate is moderate.*

BODY-126 *North Korean UAVs are expected to show minimal changes in heat and digital signatures over the next decade due to technological limitations. Despite efforts to improve drone capabilities, crashed UAVs recovered by South Korea were found to be primitive.*

BODY-127 *North Korea lacks significant drone spending compared to other countries, relying on China for technology. Their focus is more on nuclear and ICBM programs. Analyst confidence in this estimate is moderate.*

BODY-128 *North Korea is expected to enhance their cyber capabilities by 2030, making it harder to detect their malware tracking and digital imprints. Their evolving technology and faster internet service will contribute to this advancement.*

BODY-129 *North Korea conducts cyberattacks with identifiable signatures, evolving to reduce digital imprint. US tracks hackers using malware files. North Korea's cyber program poses evolving threat to US by 2030. Kim Jong-Un values cyber warfare. Thousands of capable hackers suspected.*

BODY-130 *North Korea expanded cyber attacks to Turkey from Namibia; moderate confidence in estimate due to reliable sources and structured method used by inexperienced analyst.*

BODY-131 *North Korea unlikely to modernize GPS jamming in 10 years due to economic limitations, will likely purchase older equipment from China and Russia.*

BODY-132 *North Korea is a challenging intelligence target due to difficulty in collecting information. They are likely to invest in Electronic Warfare capabilities, but it will have minimal impact on their signature in the next 10 years. Analytic confidence is moderate.*

BODY-133 *North Korea's ICBM program likely to be easier to detect by 2035 due to enhanced thermal, speed, physical, and visual signatures. Materials from Russia and China, as well as black market purchases, contribute to program advancement despite economic sanctions.*

BODY-134 *North Korea funds its ICBM program through cyberattacks, generating billions for weapons programs. Analysts predict challenging growth by 2030, with potential to target the US. Confidence in estimate is moderate, based on reliable sources and structured analysis.*

BODY-135 *The page is about unclassified information.*

BODY-136 *Advances in AI, ML, and sensor technology will enhance autonomous machines in the next 10-20 years, with lower costs and improved safety. Countries and industries are pushing for progress, with developments in supercomputers and 5G networks playing a key role.*

BODY-137 *The page discusses the competitive market for mobile robots, drones, and autonomous vehicles in delivery and warehousing, as well as China's efforts to grow its domestic chip industry. It also touches on the slow innovation in autonomous systems and the export of autonomous platforms by China to other countries.*

BODY-138 *Adversaries unlikely to use Artificial Humans for disinformation due to current limitations in AI capabilities and processing speeds. Samsung and UIMC developing AHs for various tasks, but widespread use in warfare unlikely in near future.*

BODY-139 *Advancements in AI and AH technology allow for lifelike virtual individuals to be projected anywhere, with concerns about video manipulation and the need for software to detect AH videos. Analytic confidence in uncovering adversarial AH development is moderate.*

BODY-140 *Robots and AI may enhance biological weapons development through 3D printing in the next decade, with countries like France and Germany contributing to weapons proliferation. Additive manufacturing, AI, and robots will drive the development of biological weapons, increasing productivity and optimizing transmissibility.*

BODY-141 *Advancements in gene editing, machine learning, and 3D printing may facilitate the development of more effective biological weapons, posing a threat to global security.*

BODY-142 *AI can enhance the printing process by reducing errors and improving performance, especially when combined with 3D printing. Expertise in AI for printing can be found globally. Analytic confidence in this estimate is moderate.*

BODY-143 *Arms trade likely to increase over next 10 years, with France and Germany among top exporters. Global arms industry sales on the rise, US dominates market. French exports expected to rise due to focus on robotic and AI development.*

BODY-144 *France and Germany are developing advanced military technology, with France focusing on fighter jet and naval exports, while Germany targets European states and volatile regimes. The French-German fighter program is progressing, with potential Spanish involvement. Future military sales are expected to increase due to technological advancements.*

BODY-145 *Spain's defense industry declined but arms exports likely to increase in next 10 years due to increased purchases from various countries. Top arms customers include Australia, Singapore, and Turkey. New government officials are improving arms import and export capabilities.*

BODY-146 *Spain signed an agreement for a strategic defense project with Germany and France, led by Minister of Defense Margarita Robles. Major projects include new frigates, submarines, satellites, helicopters, and aircraft modernization, positioning Spain for increased arms trade by 2030. Moderate analytic confidence in estimate.*

BODY-147 *Israel's arms trade is expected to increase over the next 10 years, driven by new customers in Asia and Europe. Despite opposition from human rights organizations, Israel's arms exports support its economy and national security. Top customers include India, Azerbaijan, and Vietnam.*

BODY-148 *Israel's arms trade has seen significant growth, with India being a major customer. The COVID-19 crisis may impact future sales, but Israel's advanced technology and diplomatic relationships continue to drive success. Opposition to Israel's arms trade persists, but it remains a key player in the global market.*

BODY-149 *Israel's arms exports face criticism for violating human rights, but continue to thrive due to international trends and legislation. Reports suggest a significant increase in arms trade, with potential new agreements with countries seeking U.S. weapons.*

BODY-150 *South Korea's arms trade is likely to increase over the next 10 years due to the North Korean threat. South Korea is moving towards self-reliance in weapon production to reduce dependency on the US.*

BODY-151 *South Korea is heavily armed and plans to increase domestic arms production, with a focus on defense technology and a significant budget increase for weapons improvements, including an aircraft carrier and advanced fighter jets.*

BODY-152 *South Korea faces increased demands for financial contributions from the US and companies withhold information while seeking high payouts. South Korea plans to build an aircraft carrier based on amphibious warfare ships.*

BODY-153 *UK arms exports may have modest growth in the next 10 years due to modernization efforts to counter Russia and meet NATO obligations. Top customers include Saudi Arabia, Oman, and the US. Defense expenditures are expected to increase, driven by technological advancements.*

BODY-154 *UK air force and army are planning modernization efforts, including new combat aircraft, ISR equipment, unmanned aerial vehicles, and directed-energy weapons. The UK Ministry of Defence is investing in cyber operations centers and new vehicle systems. The army is adapting to a technical revolution and increasing threats in the cyber domain.*

BODY-155 *The British Army is modernizing its fleet with new vehicles, investing in research and development, and aiming to increase arms exports.*

BODY-156 *The UK government aims to strengthen defense and security industries through strategic partnerships and support from the UK Export Finance, focusing on aircraft, vessels, and missiles for future export sales.*

BODY-157 *The page is about unclassified information.*

BODY-158 *Graph showing estimated threats to the U.S. Army in next 10-20 years, including anti-satellite operations, cyber, electronic warfare, hypersonics, quantum computing, short-range strike capability, and unmanned systems.*

BODY-159 *Chart showing China's estimated future signature and threat capability, using a difficulty scale to assess US ability to detect signatures on platforms in 10-20 years.*

BODY-160 *Future signature and threat capability matrix for China, using a Difficulty Scale to assess US ability to detect signatures on platforms in 10 to 20 years.*

BODY-161 *Chart showing estimated future signature and threat capability matrix for Russia, using a difficulty scale to analyze US ability to detect signatures on platforms in 10-20 years.*

BODY-162 *Future signature and threat capability matrix for Russia, using a Difficulty Scale to assess US ability to detect signatures on platforms in 10-20 years.*

BODY-180 Comparison of future capabilities at the signature/country level, with outer rim indicating higher likelihood of difficulty in detection by the US in 10 to 20 years.

BODY-181 Comparison of global/country future capabilities and signatures in terms of detectability by the US in 10 to 20 years.

BODY-182 Comparison of global/country future capabilities and signatures in terms of difficulty to detect by the US in 10 to 20 years.

BODY-183 Comparison of global/country capabilities and signatures in the future, with outer rim indicating higher likelihood of capability and difficulty in detection by the US in 10 to 20 years.

BODY-184 Comparison of global/country capabilities at signature level, with outer rim indicating higher likelihood and difficulty of detection by US in 10-20 years.

BODY-185 Comparison of future capabilities and signatures at the global/country level, with outer rim indicating higher likelihood of capability and difficulty in detection by the US in 10 to 20 years.

BODY-186 Comparison of future capabilities and signatures, with outer rim indicating higher likelihood and difficulty of detection by the US in 10 to 20 years.

BODY-187 The page is about unclassified information.

BODY-188 The Standard Primary Source Credibility Scale, also known as "The Paul Scale," evaluates the credibility of primary sources based on factors such as track record, corroboration, expertise, trustworthiness, bias, and articulateness.

BODY-189 This page provides a tool to evaluate the reliability of electronic sources based on criteria such as content corroboration, author reputation, and site bias. It offers tips for determining credibility and assigns a trust score to each source.

BODY-190 The page explains how source reliability is indicated in citations, with links provided for further verification. It also mentions the use of the Trust Scale and Website Evaluation Worksheet for determining reliability.

BODY-191 Kesselman List of Estimative Words refines intelligence assessment probability scale, based on Kent and NIC lists, with unique phrasing and odds equivalents.

BODY-192 Estimative charts show probability of detecting signatures from 2030-2040, ranging from extremely easy to detect to much more difficult.

BODY-193 Projected advancements in key deployable technologies from 2020-2040 include sensors, computers, communications, artificial intelligence, robotics, missiles, explosives, fuels, and propulsion systems.

BODY-194 Projected advancements in key deployable technologies from 2020-2040 include improvements in projectiles, propulsion, platforms, armor, stealth, satellites, weapons, and technologies such as radio-frequency weapons, nonlethal weapons, biological and chemical weapons, particle beams, electric guns, lasers, nanomaterials, 3D printing, and human enhancement devices.

BODY-195 Terms of Reference for LTG Scott D. Berrier regarding Likely Threat Signatures & U.S. Army Sensor Technology in 2030-2040 by Team Sensing at USAWC on November 12, 2019.

BODY-196 Research project on likely future threat signatures in 2030-2040 and the sensor technology needed by the U.S. Army to detect, recognize, analyze, and target these threats. Methodology includes using open source information, breaking down the research question into sections, and utilizing various sources for information gathering.

BODY-197 *Team Sensing will use various analytic techniques to enhance data analysis and ensure quality control. Challenges include time constraints, limited funding, and lack of experience in futures forecasting.*

BODY-198 *Team has intelligence experience, access to resources, and a dedicated researcher for a project with an estimated end date of April 2020. Correspondence will be through primary contact LTC Sam Smith. Team members listed with their roles. Definitions of terms provided.*

BODY-199 *Definitions of key terms related to cyberspace, detection, full-spectrum superiority, ISR, land domain, recognition, sensors, signature, space domain, and target acquisition in military operations.*

BODY-200 *Definitions and terms used in military intelligence, including acoustic, anti-satellite, artificial intelligence, batteries, ballistic missile, bio/chemical defense, C4ISR, cyber, deception, digital, and electronic warfare support.*

BODY-201 *Electronic warfare involves the use of electromagnetic energy to control the spectrum or attack the enemy. It includes electronic attack, protection, and support. Other terms discussed include electronic intelligence, frequency, hyperspectral imagery, hypersonic speed, infrared radiation, and missile types.*

BODY-202 *Definitions of various military and technological terms, including missile types, quantum computing, lasers, meta-analysis, radiation, radio waves, smell, and software.*

BODY-203 *Definitions of various military and scientific terms related to technology, radiation, and unmanned systems, including seismic, speed, stealth, tank, thermal, thermal radiation, thermal imaging, ultraviolet, unmanned aircraft, unmanned aircraft system, visual, wavelength, and X-ray.*

Notable Passages

BODY-5 *"Based on an analysis of global modernization plans, defense spending, and the stated goals of senior leaders from China, Russia, North Korea, Iran, and other Global Arms Exporters (GAE) such as Germany, Israel, and France, it is highly likely that there will be 19 updated or new capabilities able to produce 22 types of unique technical and non-technical threat signatures in 2030-2040."*

BODY-6 *There is, however, likely a concentration of signatures, spread across multiple countries and capabilities, which, if addressed, could provide a modest strategic advantage. This concentration is depicted in figure 2 below, and illustrates four global signatures that will likely become, on average increasingly more difficult to detect based on changes or masking of the adversaries' physical (the actual physical characteristics of an object), visual (the visibly detectable physical characteristics of an object), electronic (digital and software), or other non-technical signatures (such as doctrine, tactics, policies and procedures characteristics). With moderate to high confidence, these signatures will likely be altered by Deception, EW, Cyber, Radar Absorbent Material (RAM), camouflage, or a combination thereof and/or enhanced by*

BODY-7 *China created the Strategic Support Force (SSF) in December 2015 to integrate the PLA's space, cyber, and EW domains to enable multi-domain war-fighting capabilities, to fight and win future information wars.*

BODY-8 *China's president, Xi Jinping, reinforced China's AI priority during his remarks at a 2018 Politburo study session that China must lead the global enterprise in AI core technology.*

BODY-9 *According to analysts at the Hoover Institution at Stanford University, Russia will incorporate many different Air Defense, Missile, EW, Cyber, and C4ISR radars and other sensors and systems and use Radar Absorbent Materials (RAM) and other adaptive camouflage, or active camouflage technology to reduce and conceal its physical, thermal, frequency, infrared and electronic threat signatures. As a result, Russian tank signatures will likely be lighter, smaller, quieter, and faster.*

BODY-10 *b. According to a 2018 Carnegie Endowment for International Peace publication, the Magic Kitten, the Iranian cyberespionage element, operates under the same digital cyber signature patterns and has the capabilities to inflict disruptive attacks. Russian, Chinese, and North Korean cyber experts are highly likely providing support to Iran.*

BODY-11 *Improvements in cyber, AI, and other technologies will likely provide opportunities for weapons proliferation and great power competition in the coming years.*

BODY-12 *"The significant evolution and advancement in three critical technologies, artificial intelligence, stealth technology, and emerging deception practices, are likely driving the bulk of the changes in the signatures that are becoming harder to detect."*

BODY-13 *"Advanced data processing technologies with sensors and systems will enable surveillance coverage rates necessary to confront future threats. Due to wide-band communication capabilities via satellite or terrestrial channels that provide surveillance products on-demand to units in the field, sensors, and systems in 2030-2040 will require the data storage and the tools necessary to exploit automated targeting."*

BODY-14 *"Systems of the Future. As previously discussed, it is highly likely there will not be a single system to discern all future signatures needed to identify or thwart enemy capabilities. While sensor systems will individually have weak signals, the best avenue to ensure correct identifiability or action is to integrate multiple weak signals from multiple sensors into a connected and integrated system to achieve precision."*

BODY-15 *CHINA LIKELY TO WEAPONIZE AI BY 2030; ENABLED AI SYSTEMS LIKELY DIMINISHES ELECTRONIC, DIGITAL, AND METANALYTIC SIGNATURES*

BODY-16 *"NORTH KOREA LIKELY TO ENHANCE CYBER CAPABILITY TO SUSTAIN NUCLEAR, INTERCONTINENTAL BALLISTIC MISSILE PROGRAMS" - 120*

BODY-19 *Based on an analysis of China's stated modernization guidance, defense spending priorities, and efforts pursuant to stated national goals in an evaluation of 19 capabilities against 22 possible signatures, it is highly likely that top four (4) most impactful signatures associated with Electronic (digital/software), Frequency, Radio Wave, and Doctrine & Non-technical Process signatures will both impact the most capabilities and become difficult, on average to detect in the 2030-2040 timeframe.*

BODY-20 *President Xi Jinping, in his address at the 19th National Congress of the Communist Party of China, stated, "We must place greater focus on combat, encourage innovation, build systems, increase efficacy and efficiency, and further military-civilian integration."*

BODY-21 *China created the Strategic Support Force (SSF) in December 2015 to integrate the PLA's space, cyber, and electronic warfare domains. Both the Institute for National Strategic Studies (INSS) and analysts for the Cyber Defense Review assess that the Network System Department of the SSF is likely the convergence of PLA's capabilities for cyber, electronic, and psychological warfare into a single force which likely enable it to take advantage of synergies among operations in these domains.*

BODY-22 *"It is [integrated electronic warfare (INEW)] methods that use a combination of electronic warfare and network warfare techniques to attrit and disrupt the adversary's networked information systems, while defending one's own, in order to secure information dominance over the battlefield." - Dean Cheng, Senior Research Fellow at the Heritage Foundation*

BODY-23 *China's modernization of autonomous capabilities will make frequency, radio wave, visual, and wavelength signatures harder to detect by 2030.*

BODY-24 *"With China's close ties between its government and private sector firms enable China's defense and intelligence industries to focus AI research and implementation, using China's vast collection of data, China is likely to weaponize AI to aid military with intelligence analysis, decision-making, vehicle autonomy, weaponry, and information operations."*

BODY-25 *"Due to the increasing convergence of artificial intelligence (AI), cyber, autonomous and unmanned systems, electronic warfare (EW), and space-based systems within China's military modernization efforts, China's C4ISR capabilities are likely to make signature detection more difficult across 11 of 22 analyzed signature categories by 2030."*

BODY-26 *China continues to gain C4ISR experience and capabilities augmented by equipment modernization efforts and practiced implementation of full-spectrum, integrated, multi-domain operations. EW and cyber warfare are a more common feature in PLA exercises to replicate complex electromagnetic spectrum conditions. The PLA likely gains synergy from full-spectrum, integrated network electronic warfare (INEW) tactics, including use of unmanned systems, autonomous systems, space-based reconnaissance or communication, EW systems, and the weaponized use of AI and cyber. Synergy also provides benefits from the signature reduction each modernized capability provides. It is likely that China will offset vulnerabilities with enabling technologies such as AI, cyber, stealth, deception practices, and quantum capabilities.*

BODY-27 *The analytic confidence for this estimate is moderate. Open sources were used and the reliability of the sources available on this topic were above average with several high-quality sources available for the estimate. Source content varied; however, sources tended to corroborate one another.*

BODY-28 *"Despite pursuing high-powered, ground-based EW systems (especially jammers) based on vacuum-tube technology instead of solid-state electronics which likely increases detectable RF, IR, and thermal signatures, China is likely to offset those vulnerabilities with full-spectrum, integrated network electronic warfare (INEW) tactics, including the weaponized use of AI and cyber."*

BODY-29 *China created the Strategic Support Force (SSF) in December 2015 to integrate the PLA's space, cyber, and EW domains, to enabling multi-domain war-fighting capabilities, to fight and win future informatized wars. The unit's mission reportedly is seizing and maintaining battlefield information control. The Network System Department of the SSF likely is the convergence of PLA's capabilities for cyber, electronic, and psychological warfare into a single force which likely enable it to take advantage of key synergies among operations in these domains. Merging the capabilities possibly increases a unit's signature in terms of personnel and equipment needed to gain synergies across the domains, especially towards controlling complex electromagnetic environments.*

BODY-30 *China demonstrated advancements in EW through airframe modernization efforts of the 2015 J-16D variant and the 2017 Chengdu J-20 potentially outfitted with electronic attack capabilities. China modified some H-6G bombers to an electronic warfare configuration fitted with electronic countermeasure pods that can engage in combat missions using electronic jamming, suppression, and anti-radiation measures. Additionally, China is pursuing heavy fuel engines for UAVs which potentially extends the air vehicle's endurance and payload capacity. According to the 2016 China Military Power Report, China's EW weapons include jamming equipment against multiple communication and radar systems and GPS satellite systems. EW systems are also being deployed with other sea- and air-based platforms intended for both offensive and*

BODY-31 *Without US mitigation efforts, China's holistic EW approach likely complicates US ability to simultaneously counter EW across multiple domains.*

BODY-32 *China's hypersonic weapon advancements throughout 2030-2040 are likely to make radio waves, thermal, IR, and non-technical signatures more difficult to detect.*

BODY-33 *At hypersonic speeds, the exterior temperature of a hypersonic vehicle or weapon can exceed 2,000°F, necessitating advanced materials that will protect interior electronics. Chinese scientists have developed a new heat-resistant material for hypersonic aircraft that can endure more than 3,000°C from heating caused by its passage through the atmosphere at speeds between Mach 5 and Mach 20. These special materials will make detection of thermal and IR signatures more difficult in the future.*

BODY-34 *Advances in hypersonic technologies have significant implications for national security. The MDA has been directed to develop and deploy defenses against hypersonic and cruise missile threats as soon as technologically able. In 2016, the Air Force Studies Board, part of the U.S. National Academy of Science, published a report on the threat posed to U.S. interests by high-speed maneuvering weapons systems. The report highlighted that because hypersonic weapons operate below the normal trajectories of ballistic missiles and above the speeds of non-hypersonic cruise missiles which exceed the capabilities of the United States' current missile and air defense systems and infrastructure; hypersonic weapons combine speed and maneuverability between the air and space regimes to produce significant new offensive capability that could pose a complex*

BODY-35 *The analytic confidence for this estimate is moderate. Open sources were used and the reliability of the sources available on this topic were above average with several high-quality sources available for the estimate. Source content varied; however, sources tended to corroborate one another.*

BODY-36 *"Chinese scientists claiming stealth breakthroughs is a recent example of China's efforts to modernize its various warfighting platforms and defenses, as well as answering Xi Jinping's national direction."*

BODY-37 *"We will further implement the military-civilian integration strategy, and speed up efforts to make innovations in defense-related science and technology." - Chinese Premier Li Keqiang*

BODY-38 *"China was developing stealth and camouflage materials that could shelter an aircraft or vessel from broad electromagnetic spectrum. . . a new material, which was said to be lighter, more robust and flexible than materials used by the military, could be also used on vehicles and clothing." - Business Insider India, quoting the Chinese Communist controlled tabloid, Global News*

BODY-39 *"Due to their the most aggressive and advanced autonomous military program to date, by 2030, China will reduce their autonomous system signatures of frequency, radio wave (RW), visual, and wavelength making it difficult for adversaries to detect those systems."*

BODY-40 *China's race for autonomous military power has China's large industrial engineering companies like CATIC (China National Aero-Technology Import & Export Corporation) focusing on autonomous technologies. CATIC recently created the Ziyan Blowfish (Figure11), an unmanned helicopter equipped with bombs that "autonomously performs more complex combat missions, including fixed-point timing detection, fixed-range reconnaissance, and targeted precision strikes."*

BODY-41 *"Of interest is the reported (future) ability to integrate with other autonomous systems, like drones or boats. The craft can provide China's Marines with firepower, conduct reconnaissance, or be staged for future use. The difficulty the US faces in tracking signatures of China's autonomous landing craft is its capability to transition from one domain to another sea to land and land back to sea."*

BODY-42 *Analytic confidence on this analysis is moderate. Open sources were used and the reliability of the sources available on this topic were above average with the several high-quality sources available for the estimate. Source content varied; however, sources tend to corroborate on another.*

BODY-43 *"Due to China's previously established and demonstrated laser technologies, it is likely that China will mature existing laser technology to create a mobile, ground-based laser capable of damaging the optics or sensitive components of low Earth orbit (LEO) satellites within the next 15 years."*

BODY-44 *"An intense laser strike of 300 watts per square centimeter can melt the surface of satellite optical glass and cause optics to fail which would degrade surveillance from satellites."*

BODY-45 *"With mass production of advanced and reliable UAVs, China is making an unprecedented statement in the global market and continues to demonstrate its ability as a challenging near-peer competitor."*

BODY-46 *"China's latest UAV technology and production is advancing, but not yet comparable to that of the U.S. and other near-peer technological advancements. However, new emerging small size UAVs signatures including frequency, physical, radio, and visual are likely to improve in China's future UAVs development."*

BODY-47 *"China's UAV weapon payloads are rapidly increasing in lethality and distance of travel. A recent assessment on the development of a unmanned carrier aerial vehicle (UCAV) demonstrated China's aggressive efforts to technologically surpass all competitors."*

BODY-48 *"The U.S. recently developed its first UCAV platform but is behind China's efforts due to other priorities. When comparing China and US UCAV's technology, Elsa Kania, a fellow at the Center for New American Security, said that 'the drone might even allow China to partially catch up to the U.S. in carrier-based combat.' Furthermore, a U.S. Department of Defense (DOD) annual report to Congress emphasized China's rapid UAV production and marketing where 'the Middle East and North Africa region was China's second largest regional arms market, probably because of the demand for armed UAVs – a niche are where China is one of the few supplier.'"*

BODY-49 *China has made it a national goal to acquire foreign technologies to not only advance its economy, but also to use these technologies to advance its military capabilities, and it is doing so through both licit and illicit means."*

BODY-50 *"China's latest UAVs demonstrations and sales reinforce the validity of sources."*

BODY-51 *China is likely to weaponize AI to aid military with intelligence analysis, decision-making, vehicle autonomy, weaponry, and information operations by 2030. Military systems augmented with AI likely to diminish electronic, digital, and metanalytic signatures making those systems more difficult to detect.*

BODY-52 *"Just as oil fueled the industrial age, data is fueling advances in the AI age. As The Economist stated, China is 'the Saudi Arabia of data.' The chief advantage China has over the U.S. is the access to troves of data from online commerce and social networks afforded by its scant privacy protections."*

BODY-53 *"[China's] hope is that they will have an advantage in adopting AI technology, and they will use that to build a military that is stronger than that of the United States."*

BODY-54 *By 2030, China is highly likely to employ the weaponized use of AI (see China AI article), which includes machine learning, to improve offensive cyber capabilities to increase attack successes while reducing common cyber signatures. Unclassified, open source reports indicate that adversaries have not yet used AI for offensive cyber, or possibly not used AI to its fullest potential. Theoretically, an improperly taught AI system or an AI system that learns incorrectly can result in unwanted outcomes. Additionally, AI powered cyber-attacks might reveal themselves should the number of attacks increase significantly on a system as the unsupervised AI learns, adapts, and re-engages the system's defenses. Despite some theorized shortcomings, China is investing heavily in, learning from, and*

BODY-55 *China continues to steadily improve its cyber capabilities as expressed in published, national strategies. China's 2013 Science of Strategy contends that the PLA should integrate space, cyber, and electronic warfare operations to "paralyze enemy operational systems." China's Military Strategy published in May 2015 identified cyberspace as one of four "critical security domains" and stated the government's intention to speed up the development of a cyber force to tackle "grave security threats" to China's cyber infrastructure. The Strategic Support Force (SSF) was created in December 2015 to integrate the PLA's space, cyber, and EW domains, enabling multi-domain war-fighting capabilities to fight and win future informatized wars.*

BODY-56 *The report added that there was a "high degree of collaboration" between academia, the private sector, and government and military organizations in Chinese IoT security research, claiming that China's "coercive apparatus supervises and directs the collection and release of vulnerabilities", and that "this practice of stockpiling vulnerabilities for exploitation extends to IT products sold in US markets".*

BODY-57 *"War is evolving in form towards informationized warfare, and intelligent warfare is on the horizon."*

BODY-59 *It is highly likely that three signatures—including non-technical signatures such as doctrine, electro-optical signatures, and signatures associated with software changes—will both impact the most capabilities and become harder, on average, to detect in the 2030-2040 time frame.*

BODY-60 *"Russia must put the development of digital technology equipment, artificial intelligence, drones and robot systems on its agenda in terms of Russian military construction." - Russian President Putin*

BODY-61 *"Offensive cyber is playing a greater role in conventional Russian military operations and may potentially play a role in the future in Russia's strategic deterrence framework. Although the Russian military has been slow to embrace cyber for both structural and doctrinal reasons, Russia has signaled that it intends to bolster the offensive as well as the defensive cyber capabilities of its armed forces."*

BODY-62 *"Over 68% of these capabilities will likely be difficult to detect in the future as 90 out of 176 signatures in these categories, over 50% become harder, on average, to detect in the 2030-2040 time frame across six primary signatures—including non-technical signatures such as doctrine, electro-optical signatures, and signatures associated with visual, speed, thermal and software changes."*

BODY-63 *"It is likely sustained investment in modernization efforts and military R&D is necessary for Russia to remain a credible military power in 2030-2040."*

BODY-64 *By 2025 or later, the EW Forces could emerge as a new combat arm with a pivotal role in military operations this would mean it moved from a support role to a fully-fledged combat arm.*

BODY-65 *"Russia will advance and integrate its C4ISR, EW, Cyber and Deception capabilities and tactics to capitalize on its layered air defense initiatives as the S-300, S-400 and future S-500 are most effective as part of a much wider integrated air defense system (IADS)."*

BODY-66 *"It will be significantly difficult to detect Russian Tank signatures in 2030-2040 due to reduced visual, thermal, infrared and electronic signatures."*

BODY-67 *"To analyze and target the critical Russian EW, Deception, Cyber, C4ISR, Tank, Hypersonic, Air Defense, Missile and Stealth capabilities in 2030-2040, multifunctional sensors and systems that can detect and recognize, electronic/digital, visual, non-technical, speed, physical, frequency, and magnetic signatures will likely be required."*

BODY-68 *"Despite Western sanctions, Russia laid out in strategic security documents the focus of its military modernization: To accelerate the upgrading of weapons and equipment and to commission new weapons in order to achieve the minimum goal of 70% modernized."*

BODY-69 *It is likely the majority of modernization will achieve the stated goal in 2020 and Russia will focus its future modernization on four critical capabilities: EW, Tanks, Air Defense and Missiles through a complex mix of design features, performance, mission planning, and tactics leveraging Cyber, C4ISR, Deception and Stealth technologies to reduce signatures and emissions from the probability of detection.*

BODY-70 *"Due to the Russian defense industry being overwhelmed when it comes to mass-producing quality military equipment, achieving the twin goals of high-volume production and the development of entirely new weapon systems as illustrated in Figure 32 is proving to be difficult to achieve, threatening the goal of 70% of full military modernization by 2020."*

BODY-71 *Russia has gone through transformative military reforms and modernization and Russia's yearly defense budget is somewhere between $150 and $180 billion and the state of Russia's military reforms and modernization programs is focused on conducting long-term investment in defense modernization and defense planning to advance its military capabilities. The United States spends about 30% of its defense budget on procurement and R&D, while Russia spends close to 50%.*

BODY-72 *"Due to the planned modernization, the Russian Armed Forces will likely increase its strike force and firepower capability as well as various air-defense assets and high-precision systems. Russia will continue to improve and develop new advanced weapons systems such as long-range strike systems, C4ISR, EW, Unmanned systems and air defense capabilities."*

BODY-73 *"Russia must put the development of digital technology equipment, artificial intelligence, drones and robot systems on its agenda in terms of Russian military construction." - Russian President Putin*

BODY-75 *"Russia will highly likely use high power signals in 2030-2040 emitted throughout a large range of frequencies to overload the receivers of enemy electronic systems and as a result will develop material to shield electronics from EW countermeasures and field fabric to reduce radar signatures."*

BODY-76 *Due to Russia likely improving its AI-driven military application capability Russia's Advanced Research Foundation announced it will design an AI-enabled system to sift through vast quantities of satellite imagery for faster and more accurate analysis.*

BODY-77 *The introduction of unmanned EW systems as a mean to advance an EW combat operation on to enemy territory is a stated Russia objective and will likely advance this capability by 2030. The use of UAVs in EW is likely to significantly increase as the use of quadcopters as antennae for ground-based EW systems allows the use of a non-aerodynamic antenna, thereby opening up a wide range of different configurations for jamming and direction finding with higher payloads.*

BODY-79 *"It is highly likely that future Russian tanks will significantly reduce their infrared, acoustic, and radar signatures which would indicate that in 2030-2040 Russian tanks will be able to blend in with their heat surroundings better, due to leveraging stealth technology that reduces or masks the temperature of their exhausts and the heat from weapon barrels."*

BODY-80 *The T-90M, is highly likely to be Russia's most advanced front-line tank for the next 10 years, it is a modernized version of a tank that first entered service in 1993. The Defense Minister also stated that the Russian Tank modernization would include T-72B3M tanks with the domestic sighting and observation system, T-90M Proryv-3 and T-80BVM tanks and BMP-1AM infantry fighting vehicles.*

BODY-81 *Russia in 2030-2040 is highly likely to decrease its air defense and missile physical, thermal, frequency, infrared and electronic signatures through a complex mix of design features, performance, mission planning, and tactics. It is highly likely multifunctional sensors and systems that can detect, recognize and analyze variations in doctrine, physical, thermal, frequency, infrared, deception and electronic threat signatures and emissions will be required to target Russian Air Defense, Missile, Electronic Warfare (EW), Cyber, and Command, Control, Communications, Computers (C4) Intelligence, Surveillance and Reconnaissance (ISR) (C4ISR) capabilities.*

BODY-82 *Due to the emergence of new threat signature absorbing and camouflage materials, Russia has placed increasing importance on producing these materials to reduce signatures and emissions from the probability of detection. According to analysts at the Hoover Institution at Stanford University it is highly likely that in 2030-2040 Russia will incorporate many different Air Defense, Missile, EW, Cyber, and C4ISR radars and other sensors and systems for the detection of different types of enemy targets and use Radar Absorbent Materials (RAM) and other adaptive camouflage, or active camouflage technology to reduce and conceal its physical, thermal, frequency, infrared and electronic threat signatures.*

BODY-83 *"During all these years since the unilateral US withdrawal from the ABM Treaty, we have been working intensively on advanced equipment, data transfer, information technology and arms, which allowed us to make a breakthrough in developing new models of strategic weapons."*

BODY-84 *"During all these years since the unilateral US withdrawal from the ABM Treaty, we have been working intensively on advanced equipment, data transfer, information technology and arms, which allowed us to make a breakthrough in developing new models of strategic weapons." - Russian President Putin*

BODY-86 *Russia is preparing to confront its adversaries, be it in the air, maritime, land or cyber domains. Russia is highly likely developing and deploying an increasingly capable vast array of EW systems as "force enablers and multipliers". Russia are developing the foundation for what they call a "layered defense," where and under what circumstances EW systems are paired with the early warning radars and air defense systems like Pantsir-S.*

BODY-87 *According to one of the leading electronic surveillance companies, Era, it is highly likely that sensor and system technologies that collects radio frequency sources across a significant bandwidth range with multiple remote stations and a central processing station, can detect, and pinpoint, signals to 400 kilometers and possibly beyond will be required to analyze and target future EW threat signatures. These signals can be anything from radars, data links, as well as identification friend or foe transponders, meaning even stealth aircraft are visible.*

BODY-88 *"Due to swift developments across multiple hypersonic technologies, Russia is likely to slowly increase its hypersonic capabilities in the next 10 years."*

BODY-89 *"It's a weapon of the future, capable of penetrating both existing and prospective missile defense systems," the Russian president stated in December 2019. M Due to the frequency and visibility of Russian and Chinese testing, there is a growing sense that the United States is losing the arms race in hypersonic weapons development.*

BODY-90 *"Russia recently unveiled a weapon called the Kinzhal, and the Zircon cruise missile which reportedly travels between 3,800mph and 4,600mph – five to six times the speed of sound, and puts Russia several years ahead of the United States."*

BODY-91 *Russia have reportedly partnered with India, splitting the development costs would allow both countries to add to their capabilities.*

BODY-93 *Based on Iran's modernization plan, defense spending, regional strategy, and research analysis that evaluated 19 capabilities against 22 possible signatures, it is highly likely that five signatures— doctrinal/non-technical, physical, digital, and electronic —will both impact the most capabilities and become harder, on average, to detect in the 2030-2040 time frame.*

BODY-94 *Iran's cyber digital signatures are highly likely to improve by twofold becoming difficult to detect and acquired through the acquisition of foreign sources technology. Iran will continue to threaten private and government networks worldwide within the next 10 years. Out of 19 capabilities and 22 signatures, Iran's cyber capabilities will mostly emit the future threat signatures of doctrine, physical digital, electronic, and hyperspectral topping 6% of all future Iran's treat signatures likely becoming difficult to detect. Iran will improve cyber warfare capabilities employing legitimate and illicit strategies to obtain the latest technology including black-market acquisitions, and the foreign employment of students for academic enrichment.*

BODY-95 *Iran's SRBM signatures including acoustic, frequency, thermal, and electronic will likely become more difficult to detect in the next 10 to 15 fifteen years. Key findings supporting the improvements on the SRBM include: a recent survey from the Jane's Sentinel Security Assessment reveals that Middle East regional states find Iran's ICBM propagation and testing as the top strategic concern.*

BODY-96 *It is likely that it will be significantly difficult to detect Iran UAVs signatures in 2030-2040 due to the continue production of the Iranian increase in radar cross section, acoustic, surface temperature, and electromagnetic technology prompting U.S. forces to develop and apply multispectral systems to capture the myriad of potential signatures.*

BODY-97 *Iran's first stealth fighter demonstrated a slow progression in air power and stealth defense capabilities, and proof of a highly likely ineffective stealth fighter signature reduction within the next 15 years.*

BODY-98 *"Iran cyber warfare digital signatures are highly likely to improve its anti-detection technology by twofold through the acquisition of foreign sources technology and will continue to threaten private and government networks worldwide within the next 10 years."*

BODY-99 *"Iran's recent attacks demonstrated continuous cyber warfare research and improvements. Extensive Cyber warfare research improved Iran's ability to achieve strategic objectives without employing the military instrument of national power; additionally, experts believe that Iran's cyber-attacks against the Saudi Arabia gas company Aramco served as evidence of unprecedented efforts for continuous future improvement of cyber warfare technology."*

BODY-100 *"Iran has a good appreciation for the utility of cyber as an instrument of national power...its extensive experience in covert activities helps guide its strategy and operations using cyber as a tool for coercion and force, and it has created a sophisticated organizational structure to manage cyber conflict."*

BODY-101 *Iran's future increase in the employment of UAVs will require the U.S. and allies to generate technology to identify the myriad of UAVs signatures in the battlefield. Current technology is limited in identifying UAVs signatures which could allow Iran to disrupt, elude, or completely evade military defense systems.*

BODY-102 *Strong indicators of Iran's increase of UAVs radar cross section and electromagnetic signatures are the numerous successful Iranian UAVs involved in the Syria conflict. Iran has been capable of arming its UAVs against insurgent forces in Syria.*

BODY-103 *"The future development of U.S. anti-drone systems must account for the different types of signatures to balance the economic cost of the drone employed and the detection and defensive system. For instance, a Heritage Foundation study on Iran's growing UAV threat, and a Defense News report alluded that shooting down a cheap armed drone with an expensive missile does not benefit the U.S. economically."*

BODY-104 *Under the current operational environment, sources demonstrated a high tendency for Iran to rapidly advance UAVs research and development.*

BODY-105 *"With the advanced long-term development of ballistic missiles (BM) and foreign technological support make it likely that Iran will test a successful Intercontinental Ballistic Missile (ICBM) within the next 10 years."*

BODY-106 *"Today, Iran possesses the capability to deploy ballistic missiles (and long-range artillery rockets) against its regional neighbors, as well as Israeli and U.S. forces deployed in the region. It has bolstered this capability with assistance from China, North Korea, and Russia since the 1980s. However, Iran has also sought to reduce its reliance on foreign support and is believed to have developed considerable in-country development capabilities." - Janes*

BODY-107 *Iran's space program increases the ability for the future employment of ICBMs although additional research and development would still be needed to include atmosphere reentry technology.*

BODY-108 *"President Mahmoud Ahmedinejad said it had [almost all the positive features] of the world's most sophisticated jets. He said the 'development of the Iranian nation's military power is... for deterrence and defensive purposes.'"*

BODY-109 *"Iran has flown a reverse-engineered copy of the Lockheed Martin RQ-170 Sentinel unmanned aerial vehicle (UAV) it captured from the United States in 2011, state media reported on 10 November." - Jane's Defence Weekly*

BODY-110 *"Despite the current international embargo, it is likely that Iran will successfully continue to acquire the essential materials for modernization."*

BODY-111 *"The employment of future sensors must account for offensive and defensive operations. Defensive sensors are those systems that protect the intrusion of sensors and systems capabilities. Regulus Cyber, a professional leading cyber protection expert, explained the criticality of the cyber protection sensors and systems, and considering the increase 'with sensors being used in both manned and unmanned platforms, the need for reliable sensor protection is increasing across all sectors and industries."*

BODY-112 *Iranian UAVs and missile attack in 2019 against Saudi Arabia's U.S. issued air defense equipment, expertly avoided air attack and area denial (A2/AD) sensors and systems, demonstrating the current rendering of U.S. defensive systems as ineffective.*

BODY-113 *"Although many intrusion detection systems have been developed, most systems are difficult to implement for the sensor nodes owing to limited computation resources."*

BODY-114 *"In an MDPI Sensors Research, Samaras, Et al demonstrated that, 'nowadays, c-UAV applications offer systems that comprise a multi-sensory arsenal often including electro-optical, thermal, acoustic, radar and radio frequency sensors, whose information can be fused to increase the confidence of threat's identification."*

BODY-115 *"In the next 10 to 20 years, in order to effectively detect and acquire Iranian emerging signatures, the U.S. Army will likely need to improve wide-area electro-optical sensors mounted on UAVs facilitating the conduct of ISR operations. Future Army sensors and systems must provide 'moderate-resolution persistent-surveillance electro-optical sensors that operate during day and at night over large areas to detect vehicles and people on foot.'"*

BODY-117 *It is highly likely that four signatures—physical, visual, thermal signatures, and speed changes—will both impact the most capabilities and become more difficult, on average, to detect in the 2030-2040 time frame.*

BODY-118 *"In fact, over 2% (10/418) of future North Korea threat signatures distributed over 4 of the 19 capabilities will likely become more difficult to detect between 2030 and 2040. 23% (97/418) will become moderate to routine to detect and 75% (313/418) of the signatures associated with these 19 capabilities are likely to become easier to detect in the same period."*

BODY-119 *"The U.S. Treasury Department said Friday that North Korean state-sponsored hacking groups attacked critical infrastructure, drawing illicit funds that ultimately funded the country's weapons and missile programs."*

BODY-120 *"By 2035, the next ICBM will likely have a three-stage rocket system that will have routine to moderate signatures (speed, physical, thermal, and visual) that will make it more difficult to detect."*

BODY-121 *7. Removing capabilities which have nominal signature changes from the calculations, as seen in Figure 64, will create a slight increase in the percentage of signatures that become more difficult to detect. This may indicate that the U.S. Army will likely need Measurement and Signature Intelligence (MASINT) sensors and systems that can detect, track, identify or describe the distinctive characteristics of North Korea's visual, thermal, frequency, radio waves, and speed signatures. That broad signature approach will capture the key capabilities of the country's Artificial Intelligence (AI), Electronic Warfare (EW), Long-Range and Short-Range Strikes (LRBM/SRBM), Missiles Intercontinental Ballistic Missile (ICBM), Cyber and Unmanned Vehicle Systems (UAV).*

BODY-122 *"In 2018, North Korea developed the Reconnaissance General Bureau (RGB), North Korea's equivalent to the CIA, [and] has trained the world's greatest bank-robbing teams."*

BODY-123 *"The world will have to learn to live with North Korea's capability to target the United States with nuclear weapons." - Jeffrey Lewis, Middlebury Institute of Strategic Studies*

BODY-124 *It is likely that North Korea will modernize their long and short ballistic missile strike capability by modifying their launch thrust ratio that will enhance their speed, and increasing their thermal signature making these signatures more difficult to detect by 2030. (Figure 66) M Kim Jung-Un limits Korea's funds that are not dedicated toward his missile program due to economic challenges. Instead the country are likely to modernize the entire missiles program by purchasing older long and short-range ballistic missiles (LR/SRBM) from Iran, Russia or China to enhance their capability. L North Korea's long and short strikes will have signatures that support electro-optical terminal guidance system which can be controlled from land or by aircraft. Their long and short-range strike signature will have the*

BODY-125 *"Despite North Korea's successful launched space satellites that demonstrated their countries technological capabilities which is required as they plan to modify their long and short-range strikes ballistic missiles by 2030. If North Korea is capable of placing additional satellites in low orbit this would increase strategic defensive and offensive capabilities, and the ability to interrupt satellite communications, but the satellite launching platform signature will likely mirror the LR/SR launch."*

BODY-126 *It is likely that North Korean UAV heat and digital signature will change at an incremental rather than dramatic pace by 2030. It is likely North Korean EW/UAV signatures will be similar in their thermal, speed, visual, electronics (digital/software) indicating that in 2030 the US will likely detect similar UAV signature as they do now.*

BODY-127 *"China has surpassed the United States in UAV technology, thereby making North Korea a stronger adversary in the UAV platform, given its close relationship with China. North Korea is not considered an expert in this field and appears to rely on China, Russia, Pakistan, and the black market to obtain high end UAV technology. However, it is highly unlikely that North Korea would spend billions of dollars on a UAV program but is likely to spend billions on their nuclear and ICBM program."*

BODY-128 *Despite North Korea's gross domestic product (GDP) $30 billion, the country is evolving their techniques of malware deception, digital imprints, and firmware by leveraging their cyber program.*

BODY-129 *"North Korea's cyber program continues to evolve as a threat to the US and will create a greater challenge to track by 2030. North Korea's cyber program has proven they are capable of penetrating the US cyber program, creating mayhem wherever North Korea decides to attack."*

BODY-130 *In February 2020, North Korea expanded their targets to companies in Turkey, operating from a block of internet addresses traced to Namibia, one of the few countries that still maintains friendly relations with Pyongyang.*

BODY-131 *"Due to North Korea's limited economic capabilities and, their $30B gross domestic product of which 60% is spent toward their Nuclear and Intercontinental Ballistic Missile (ICBM) program, it is highly unlikely they will spend additional funds to improve their GPS jamming capability."*

BODY-132 *"North Korea is a low-level threat to the United States in Northeast Asia and but is our hardest intelligence collection target due to how hard the information is to collect." - LTG Robert Ashley*

BODY-133 *Despite North Korean economic sanctions and a low Gross Domestic Product (GDP), North Korea will likely enhance their Intercontinental Ballistic Missile (ICBM) program by modernizing their thermal, speed, physical program likely easier to detect by 2035.*

BODY-134 *"In just the past few years, RGB hackers have struck more than 100 banks and cryptocurrency exchanges around the world, pilfering more than $650 million. Overall, North Korea has generated an estimated $2 billion for its weapons of mass destruction programs using 'widespread and increasingly sophisticated' cyberattacks to steal from banks and cryptocurrency exchanges."*

BODY-136 *"Cost reductions in sensors, actuators, radar, lidar and camera systems, as well as advances in multi-sensor integration through sensor fusion, improve depth detection for safe and automated motion and bring autonomous things closer to reality."*

BODY-137 *"The market for mobile robots, drones, and autonomous vehicles in delivery and warehousing is likely to reach a staggering $81 and $290 Billion in 2030 and 2040, respectively."*

BODY-138 *"Samsung Technology and Advanced Research Labs sub-company Neon debuted its AHs project at Consumer Electronics Show (CES) 2020. Neon said their design of AHs are programable and able to have real conversations and behave like humans. The AHs are able to 'form memories and learn new skills, but don't have a physical embodiment.' Designers can personalize AH to be 'goal-oriented' in completing tasks or personalized to assist as teachers, healthcare providers, spokespeople, or television anchors."*

BODY-139 *"Video manipulation and deception have long been possible, but advances in machine learning have made it easy to automatically capture a person's likeness and stitch it onto someone else." - Will Knight, Wired Magazine*

BODY-140 *"The increased use of robots in laboratories could lead to significant gains in productivity during the design-build-test cycle of biological weapons, while artificial intelligence could be used to find new ways to optimize the transmissibility or virulence of a biological agent."*

BODY-141 *"In the future, genomic data, gene editing tools such as [clustered regularly interspaced short palindromic repeats] CRISPR, and machine learning tools may assist nefarious actors interested in developing more effective biological weapons."*

BODY-142 *"AI can definitely assist the printing process and prevent errors, as it is often related to expressions such as machine learning, neural networks, automation or artificial vision. When combined with 3D printing, it could increase performance and reduce the risk of errors by facilitating automatic production." - Ashraf Abdelhak*

BODY-143 *"Arms sales are no new matter to the world stage as conflicts continue throughout the world and regimes attempt to reign control over their country masses. The global arms industry sales are on the rise."*

BODY-144 *"As of 2018 – 2019, France's recent jump and future military sales is due to its fighter jet exports to Egypt, India, and Qatar while its naval exports have been to Brazil, India, Malaysia, Belgium, Netherlands, and the United Arab Emirates (UAE)."*

BODY-145 *"Despite the previous decline, Spain, in 2018, elected new government officials and now are aggressively improving their arms import and export capability. Part of the New Modernization plan implied generating closer institutional relations with the purpose of moving forward that will allow for greater export as well as imports."*

BODY-146 *"The Spanish defense industry faces the challenge of actively participating in the consolidation of the European defense market, where further internationalization opportunities will arise. The pace at which a single European defense market is being created must be changed so that those companies that are currently less accustomed to working in this environment get used to it." - Ángel Olivares, Secretary of State for Defence*

BODY-147 *"Despite the large efforts from human rights organizations to counter Israel's arms export, Israel continues to employ groundbreaking strategies to increase its arms trade that supports its domestic economy and national security interests."*

BODY-148 *"With an unprecedented $9.2 billion record arms trade sales in 2017, a major portion of Israel's profits were from India's purchases. One of the top customers of* *arms trade for Israel continues to be India with diplomatic relationship stronger than ever, and even in the midst of the COVID-19, and according to a statement released by the Indian government, 'Israel will supply the Indian military with 16,479 Negev light machine guns, despite growing concerns over the health emergency facing the country of 1.3 billion people.'"*

BODY-149 *"Despite the large efforts from human rights organizations to counter Israel's arms export, Israel continues to gain remarkable profits while, according to Toi, from The Times of Israel, in March 2019 'the Defense Minister believes the increase in sales stems from several international trends, one being is an increase in defense budgets of NATO countries.'"*

BODY-150 *"In the long run, those weapons are going to be exported around the world."*

BODY-151 *"A key motive for South Korea to become an independent weapons developer, producer, and exporter of military arms stems from the frustrations with the US."*

BODY-152 *"companies do not share the latest development while demanding a huge payout from South Korea to help cover their research and development costs."*

BODY-153 *"Driven by the need to counter a resurgent Russia, the UK's defense expenditures are anticipated to increase from US$50.3 billion in 2019 to US$52.6 billion in 2020. Technological advancement will upsurge the demand for the replacement of ageing equipment and the procurement of new products, will likely propel the defense market in the coming years."*

BODY-154 *"Secure borders, or living on an island, are no guarantees against the corrosive and intrusive effects of disinformation, subversion and cyber." - Mark Carleton-Smith*

BODY-155 *"The British Army, therefore, has a modernization challenge. We are seeking to invest in new and novel technologies whilst simultaneously attempting to modernize an ageing and increasingly obsolete fleet."*

BODY-156 *"Our relationship with industry is crucial to maintaining the UK's position as a Tier 1 military power." - UK Defence Secretary Ben Wallace*

BODY-158 *The most likely threats posed will be Anti-satellite operations, Cyber, Electronic Warfare, Hypersonic's, Quantum Computing, Short-range strike capability, and unmanned systems.*

BODY-168 *Overall Chinese Threat the closer to the outer rim the more likely their capability and signature will be more difficult to detect by the US in 10 to 20 Years.*

BODY-169 *Overall Iranian Threat the closer to the outer rim the more likely their capability and signature will be more difficult to detect by the US in 10 to 20 Years.*

BODY-170 *Overall Commercial and Other Country Threat's the closer to the outer rim the more likely their capability and signature will be more difficult to detect by the US in 10 to 20 Years.*

BODY-172 *Outer rim represents higher likeliness of capability and signature being more difficult to detect by the US in 10 to 20 Years*

BODY-173 *Outer rim represents higher likeliness of capability and signature being more difficult to detect by the US in 10 to 20 Years*

BODY-175 *Outer rim represents higher likeliness of capability and signature being more difficult to detect by the US in 10 to 20 Years*

BODY-176 *Outer rim represents higher likeliness of capability and signature being more difficult to detect by the US in 10 to 20 Years.*

BODY-177 *Outer rim represents higher likeliness of capability and signature being more difficult to detect by the US in 10 to 20 Years*

BODY-178 *Outer rim represents higher likeliness of capability and signature being more difficult to detect by the US in 10 to 20 Years*

BODY-179 *Outer rim represents higher likeliness of capability and signature being more difficult to detect by the US in 10 to 20 Years*

BODY-194 *"Finally, technology areas in which revolutionary advances occur will be able to accomplish important battlefield tasks that they cannot now even attempt."*

BODY-197 *The team will apply appropriate structured analytic techniques and methodologies that provide insight into the collected data and enhance the rigor of analysis. Techniques and methodologies will be selected based on their relative best fit with the requirement and the types of data available, and may include, but not limited to: Mind mapping, Nominal method technique, Red team analysis, Alternative futures analysis, Multi-criteria analysis, Science of estimation, Forecasting, Contrarian techniques, Diagnostic techniques, Imaginative thinking techniques.*

BODY-199 *Full-spectrum superiority — The cumulative effect of dominance in the air, land, maritime, and space domains; electromagnetic spectrum; and information environment (which includes cyberspace) that permits the conduct of operations without effective opposition or prohibitive interference. (JP 3-0)*

BODY-202 *"The term 'quantum supremacy' refers to the ability of a quantum computer to perform tasks beyond the capability of today's most powerful conventional supercomputers."*

LIKELY THREAT SIGNATURES 2030 - 2040

TEAM SEN5ING

TOMORROW'S WORK TODAY

GROUND SENSING ANALYSIS for WARFARE in 10 to 20 YEARS

Ashraf S. Abdelhak / Jerry A. Brown / Rafael A. Duran / Russell V. Hoff / Samuel P. Smith

A special THANK YOU to *Thomas G. Moss* of the United States Army Heritage and Education Center College Library. Your dedication to supporting our success and helping us become better researchers will carry us for years to come!

About This Document

The United States (US) Army War College (AWC) Futures Seminar Team Sensing prepared this document for the Deputy Chief of Staff of the Army, G2 (Intelligence). The research, analysis, and production of this product occurred over a twenty-eight-week time frame from October 2019 to April 2020 at US AWC. The team consisted of five Department of Defense civilian and active duty members: Ashraf Abdelhak, Jerry Brown, Rafael Duran, Russell Hoff, Samuel Smith.

Requirement

This product consists of multiple versions to convey the team's analysis, research, and key findings: an electronic PDF version of the report, a limited number of hard copy versions of the report, a PowerPoint presentation (briefed on 7 April 2020), and a limited number of enlarged, hard copy maps and tables. All shapefiles and spreadsheets containing raw data will be transmitted to Army G2.

What are likely future threat signatures in 2030-2040? What sensors and systems will the U.S. Army likely need in order to detect, recognize, analyze, and target future threat signatures?

Analytic Confidence

Analytic confidence is determined through the reliability and corroboration of sources used, the use of structured analytical techniques, the analysts' subject matter expertise, level of teamwork, the constraints, and question complexity. The Peterson list with the Friedman corollaries, the guide used to determine analytic confidence in this report, can be found in Annex 1. Unless otherwise noted, each Sources Analytic Confidence is *moderate*.

Words of Estimative Probability

Analysts leveraged the Kesselman List of Estimative Words, found in Annex 4, as their Words of Estimative Probability (WEP) and sliding scale to determine the likelihood of a capability's future threat, specifically in 10 – 20 years. Unless otherwise noted the estimate should be determined as *likely*.

Source Reliability

Source reliability is noted at the end of each citation as low [L], moderate [M], or high [H]. The citation is hyperlinked to the source, unless the source is a paid subscription; in that instance a footnote is provided at the end of each writing illustrating the source for credibility. Source reliability is determined using the Trust Scale and Website Evaluation Worksheet found in Annex 2.

Analyst Contact Information

Ashraf Abdelhak

ashraf.abdelhak@us.af.mil

Jerry Brown

jerry.brown2.mil@mail.mil

Rafael Duran

rafael.duranmariot.mil@mail.mil

Russell Hoff

russell.v.hoff.mil@mail.mil

Samuel Smith

samuel.p.smith.mil@mail.mil

Global Threat Capabilities, Signatures and Sensor Key Findings

What are likely future threat signatures in 2030-2040?

1. Based on an analysis of global modernization plans, defense spending, and the stated goals of senior leaders from China, Russia, North Korea, Iran, and other Global Arms Exporters (GAE) such as Germany, Israel, and France (see Figure 1 below), it is highly likely that there will be 19 updated or new capabilities able to produce 22 types of unique technical and non-technical threat signatures in 2030-2040. Across all countries and capabilities, after examining 418 possibilities, it is highly likely that, globally, no single technical signature emerges as a priority candidate for exploitation by some future combination of sensors and systems.

Estimated Signature and Capability Threat Matrix

	Cyber	Electronic Warfare	Deception	Artificial Intelligence	Short-Range Strike	C4ISR	Missiles	Long-Range Strike	Unmanned Systems	Air Defense	Anti-Satellite	Autonomous	Quantum	Chemical	Batteries	3D Printing	Hypersonic	Stealth	Tanks
Physical	4	4	3	3	4	3	3	3	3	3	3	3	3	3	3	3	3	3	3
Visual	3	4	4	3	3	3	3	3	3	2	3	3	3	3	3	3	3	3	3
Doctrine and other Non-Technical Process	4	3	4	3	3	4	3	3	3	3	3	3	3	3	3	3	3	3	3
Electronic (Digital/Software)	4	4	4	4	3	3	3	3	3	3	3	3	3	3	3	3	3	3	3
Speed	4	3	3	3	3	3	3	3	3	3	3	3	3	3	3	3	3	3	3
Frequency	4	4	3	3	3	3	3	3	4	3	3	3	3	3	3	3	3	3	3
Radio waves	3	4	3	3	3	3	3	3	3	3	3	3	3	3	3	3	3	3	3
Digital	4	3	3	3	3	3	3	3	3	3	3	3	3	3	3	3	3	3	3
Thermal	4	3	3	3	3	3	3	3	3	3	3	3	3	3	3	3	3	3	3
Wavelength	3	4	3	3	3	3	3	3	3	3	3	3	3	3	3	3	3	3	3
Radiation	3	3	3	3	3	3	3	3	3	3	3	3	3	3	3	3	3	3	3
Acoustic	3	3	3	3	3	3	3	3	3	3	3	3	3	3	3	3	3	3	3
Hyperspectral	4	3	3	3	3	3	3	3	3	3	3	3	3	3	3	3	3	3	3
Magnetic	3	3	3	3	3	3	3	3	3	3	3	3	3	3	3	3	3	3	3
Metanalytic	3	3	3	3	3	3	3	3	3	3	3	3	3	3	3	3	3	3	3
Laser	3	3	3	3	3	3	3	3	3	3	3	3	3	3	3	3	3	3	3
Smell	3	3	3	3	3	3	3	3	3	3	3	3	3	3	3	3	3	3	3
Ultra Violet	3	3	3	3	3	3	3	3	3	3	3	3	3	3	3	3	3	3	3
x-Ray	3	3	3	3	3	3	3	3	3	3	3	3	3	3	3	3	3	3	3
Seismic	3	3	3	3	3	3	3	3	3	3	3	3	3	3	3	3	3	3	2
Infrared	3	3	2	3	3	2	3	3	3	3	3	3	3	3	3	3	3	3	3
Bio/Chemical	3	3	3	3	3	3	3	3	3	3	2	3	3	3	3	2	3	3	3

5 Much More Difficult to Detect	4 More Difficult to Detect	3 Moderate to Routine Improvement	2 Easier to Detect	1 Extremely Easier to Detect

Figure 1 Estimated Signature & Capability Threat Matrix illustrates the distribution of signature detection difficulty on a scale of 1-5 across 19 Capabilities and 22 Signatures from China, Russia, North Korea, Iran, and other Global Arms Exporters.

2. On the other hand, non-technical signatures, such as doctrine and order of battle, are likely to change several times between now and 2040. It is highly likely that technological advances and new capabilities will drive technical and non-technical signatures. Having sensors and systems in place to track these non-technical signatures likely provides some of the earliest possible warnings for changes in technical signatures.

3. There is, however, likely a concentration of signatures, spread across multiple countries and capabilities, which, if addressed, could provide a modest strategic advantage. This concentration is depicted in figure 2 below, and illustrates four global signatures that will likely become, on average increasingly more difficult to detect based on changes or masking of the adversaries' physical (the actual physical characteristics of an object), visual (the visibly detectable physical characteristics of an object), electronic (digital and software), or other non-technical signatures (such as doctrine, tactics, policies and procedures characteristics). With moderate to high confidence, these signatures will likely be altered by Deception, EW, Cyber, Radar Absorbent Material (RAM), camouflage,

					Estimated Signature and Capability Threat Matrix				
	Cyber	Electronic Warfare	Deception	Artificial Intelligence	Short-Range Strike	C4ISR	Missiles	Long-Range Strike	Unmanned Systems
Physical	4	4	3	3	4	3	3	3	3
Visual	3	4	4	3	3	3	3	3	3
Doctrine and other Non-Technical Process	4	3	4	3	3	4	3	3	3
Electronic (Digital/Software)	4	4	4	4	3	3	3	3	3

Figure 2 illustrates the concentration of signature difficulty across 19 capabilities and 22 signatures resulting in four critical signatures across nine capabilities.

or a combination thereof and/or enhanced by the use of AI and Quantum.

4. These global signatures are, in turn, as illustrated in figure 3, spread across a variety of capabilities or weapons systems. Four key threat signatures across seven significant capabilities: Cyber, EW, AI, Missiles (Long-Short Range Strike), Deception, and C4ISR are highly likely to be the most difficult to detect in 2030-2040.

5. Various, country-specific capabilities and signatures may either be easier or more difficult to detect depending on the signature, capability, country, or combination of the three. The below information is relevant to specific countries posed as the current and future highest threats to US National Security:

a. **China**. It is *highly likely* China's top four most impactful signatures will be associated with electronic (digital/software), frequency, radio wave, and doctrinal or other non-technical processes. The top four Chinese capabilities that affect these signatures will be C4ISR, EW, Hypersonic, and Autonomous Systems, see figure 3.

Estimated Signature and Capability Threat Matrix - China

	C4ISR	Electronic Warfare	Hypersonic	Stealth	Autonomous	Anti-Satellite	Unmanned Systems	Artificial Intelligence	Cyber	Quantum	Deception
Frequency	4	4	4	4	4	3	4	3	3	3	3
Radio waves	4	4	4	4	4	3	4	3	3	3	3
Physical	4	4	4	3	3	4	4	3	3	3	4
Electronic (Digital/Software)	4	4	3	3	3	3	3	4	4	4	3
Doctrine & Other Non-Technical Process	4	2	4	4	3	4	3	3	3	4	4
Digital	4	4	3	3	3	3	3	4	4	4	3
Visual	4	4	3	3	4	3	4	3	3	3	4
Wavelength	4	4	3	4	4	3	3	3	3	3	3
Metanalytic	3	4	3	3	3	3	3	4	4	3	3
Radiation	4	4	3	3	3	3	3	3	3	3	3
Speed	4	3	3	3	4	3	3	3	3	3	3
Hyperspectral	4	3	3	3	3	3	3	3	3	3	3
Thermal	2	2	4	4	3	4	3	3	3	3	3
Laser	3	3	3	3	3	4	3	3	3	3	3
Infrared	2	2	4	4	3	4	3	3	3	3	3
Acoustic	3	3	3	3	3	3	3	3	3	3	3
Bio/Chemical	3	3	3	3	3	3	3	3	3	3	3
Magnetic	3	3	3	3	3	3					
Seismic	3	3	3	3	3	3					
Smell	3	3	3	3	3	3	3	3	3	3	3
Ultra Violet	3	3	3	3	3	3	3	3	3	3	3
X-Ray	3	3	3	3	3	3	3	3	3	3	3

Figure 3 illustrating the top four Chinese Capabilities which will most likely be developed using top four signatures

5 Much More Difficult to Detect	4 More Difficult to Detect	3 Moderate to Routine Improvement	2 Easier to Detect	1 Extremely Easier to Detect

To view, complete findings see page 18. Key facts summarized below.

i) In fact, 15% (62/418) of future Chinese threat signatures, distributed over 12 of the 19 threat capabilities, will likely become more difficult to detect between 2030 and 2040. China's C4ISR is likely to see the most upgrades, and consequently, the most negative overall change in the capability's signature profile, hence more challenging to detect. In fact, in 11 of the 22 signatures examined, China's C4ISR capability is likely to get more difficult to detect over the next 20 years. These changes are primarily due to forecasted improvements in enabling technologies and capabilities such as Cyber and Artificial Intelligence.

ii) China created the Strategic Support Force (SSF) in December 2015 to integrate the PLA's space, cyber, and EW domains to enable multi-domain war-fighting capabilities, to fight and win future information wars.

iii) According to the 2016 China Military Power Report, China's EW weapons include "jamming equipment against multiple communication and radar systems and GPS satellite systems. EW systems are also being deployed with other sea- and air-based platforms intended for both offensive and defensive operations."

iv) China's president, Xi Jinping, reinforced China's AI priority during his remarks at a 2018 Politburo study session that China must lead the global enterprise in AI core technology.

b. **Russia**. It is *highly likely* Russia's most impactful signatures will be associated with doctrinal or other non-technical processes, electro-optical, radio waves, speed, and electronic digital/software changes. The top four Russian capabilities that affect these signatures will be EW, deception practices, Cyber, and C4ISR, see figure 4 below. The main facts are listed below, to view the complete Russia country findings see page 54.

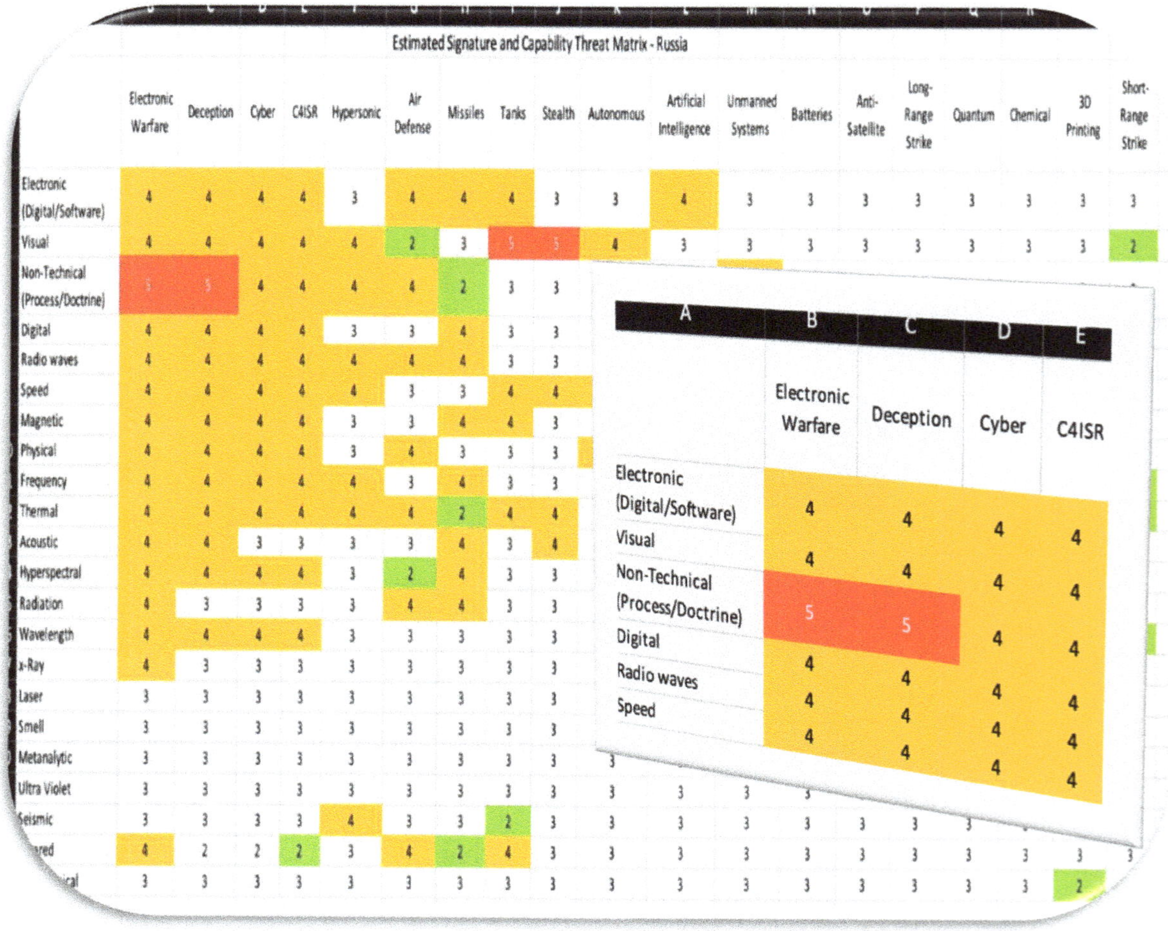

Figure 5 Russia Estimated Signature and Threat Capability Matrix

5 Much More Difficult to Detect	4 More Difficult to Detect	3 Moderate to Routine Improvement	2 Easier to Detect	1 Extremely Easier to Detect

i. The Centre for Economics and Foreign Policy Studies reported that in five to ten years, technological advancements in EW, Cyber, Deception, and C4ISR (AI) capabilities will become more advanced and difficult to detect.

ii. According to Russia's State Armaments Program for 2018-2027, Russia will invest 19 trillion rubles (about 306 billion US dollars) in the next 10 years in defense procurement and equipment upgrades including high-precision weapons for air, sea, and land battle – including hypersonic missiles – unmanned air strike complexes, individual equipment for service members and advanced reconnaissance, communication and electronic warfare systems. Russia will likely focus its modernization across 40% of its primary capabilities. See Chart 5, Aggregate Reduction 1. The future threat capabilities that will be difficult to detect are EW, Missile, Air Defense, and Tanks by leveraging Cyber, C4ISR, Deception and Stealth enabling technologies. See Figure 6, Aggregate Reduction 2. Over 68% of these capabilities will likely be more difficult to detect in the future, as 90 out of 176 signatures in these categories, over 50% of the signatures likely becoming harder, on average, to detect in the 2030-2040 time frame across six primary signatures (non-technical signatures such as doctrine, electro-optical signatures, and signatures associated with visual, speed, thermal and software changes).

iii. According to Roger McDermott, a senior research fellow in Eurasian Military Studies at King's College in London, Russian EW is found throughout every arm and branch of service, making it nearly impossible to avoid. To further demonstrate Russia's likely future capability, based on a report by the Republic of Estonia Ministry of Defense, Russia's interest in and use of EW is part of a comprehensive effort by Moscow to adopt and strengthen its network-centric capability, which focuses upon C4ISR integration. Russia is already fielding automated command and control (C2) systems that are feeding into EW capability.

iv. According to analysts at the Hoover Institution at Stanford University, Russia will incorporate many different Air Defense, Missile, EW, Cyber, and C4ISR radars and other sensors and systems and use Radar Absorbent Materials (RAM) and other adaptive camouflage, or active camouflage technology to reduce and conceal its physical, thermal, frequency, infrared and electronic threat signatures. As a result, Russian tank signatures will likely be lighter, smaller, quieter, and faster.

c. **Iran**. It is *highly likely* Iran's top five most impactful signatures will be associated with doctrinal or other non-technical processes, physical, frequency, digital, and electronic (digital/software). The top four Iranian capabilities that affect these signatures will be Cyber, Unmanned Systems (UMS), Deception practices, and

Missiles. See Figure 5 on the next page and click here to view complete Iran analysis and findings on page 87.

Estimated Signature and Capability Threat Matrix - Iran

	Cyber	Unmanned Systems	Electronic Warfare	Deception	Short-Range Strike	Air Defense	Chemical	Artificial Intelligence	Batteries	Quantum	3D Printing	Anti-Satellite	Long-Range Strike	Missiles
Frequency	4	4	4	3	4	3	[illegible]	[illegible]	3	3	3	3	3	3
Electronic (Digital/Software)	4	4	3	3	4	3	[illegible]	[illegible]	[illegible]	[illegible]	[illegible]	[illegible]	[illegible]	[illegible]
Acoustic	3	4	4	4	2	3	[illegible]	[illegible]	[illegible]	[illegible]	[illegible]	[illegible]	[illegible]	[illegible]
Doctrine and other Non-Technical Process	4	3	3	4	3	3	[illegible]	[illegible]	[illegible]	[illegible]	[illegible]	[illegible]	[illegible]	[illegible]
Physical	4	3	3	3	4	3	[illegible]	[illegible]	[illegible]	[illegible]	[illegible]	[illegible]	[illegible]	[illegible]
Thermal	4	3	3	3	4	3	[illegible]	[illegible]	[illegible]	[illegible]	[illegible]	[illegible]	[illegible]	[illegible]
Hyperspectral	4	3	3	3	3	3	[illegible]	[illegible]	[illegible]	[illegible]	[illegible]	[illegible]	[illegible]	[illegible]
Radiation	3	4	3	3	3	3	[illegible]	[illegible]	[illegible]	[illegible]	[illegible]	[illegible]	[illegible]	[illegible]
Speed	4	2	3	4	3	3	[illegible]	[illegible]	[illegible]	[illegible]	[illegible]	[illegible]	[illegible]	[illegible]
Laser	3	3	3	3	3	3	[illegible]	[illegible]	[illegible]	[illegible]	[illegible]	[illegible]	[illegible]	[illegible]
Magnetic	3	3	4	3	2	3	[illegible]	[illegible]	[illegible]	[illegible]	[illegible]	[illegible]	[illegible]	[illegible]
Metanalytic	3	3	3	3	3	3	[illegible]	[illegible]	[illegible]	[illegible]	[illegible]	[illegible]	[illegible]	[illegible]
Seismic	3	3	3	3	3	3	[illegible]	[illegible]	[illegible]	[illegible]	[illegible]	[illegible]	[illegible]	[illegible]
Smell	3	3	3	3	3	3	[illegible]	[illegible]	[illegible]	[illegible]	[illegible]	[illegible]	[illegible]	[illegible]
Ultra Violet	3	3	3	3	3	3	[illegible]	[illegible]	[illegible]	[illegible]	[illegible]	[illegible]	[illegible]	[illegible]
Visual	3	3	3	4	2	3	[illegible]	[illegible]	[illegible]	[illegible]	[illegible]	[illegible]	[illegible]	[illegible]
Wavelength	3	3	4	3	2	3	[illegible]	[illegible]	[illegible]	[illegible]	[illegible]	[illegible]	[illegible]	[illegible]
x-Ray	3	3	3	3	3	3	3	3	[illegible]	[illegible]	[illegible]	[illegible]	[illegible]	[illegible]
Digital	4	4	3	4	3	3	3	3	3	3	[illegible]	[illegible]	[illegible]	[illegible]
Infrared	2	4	3	2	3	3	3	3	3	3	3	3	[illegible]	[illegible]
Radio waves	3	3	3	2	3	3	3	3	3	3	3	3	3	3
Bio/Chemical	2	4	2	2	3	3	3	2	2	2	2	1	2	2

Figure 5 Iran's Top Signature and Capability Threats in the next 10 to 20 years

5 Much More Difficult to Detect	4 More Difficult to Detect	3 Moderate to Routine Improvement	2 Easier to Detect	1 Extremely Easier to Detect

a. A Jane's 2020 Iran Executive Summary report highlighted the creation of the Iranian Cyber Army as a critical component of government-wide Cyber Modernization effort, increasing offensive and defensive capabilities.

b. According to a 2018 Carnegie Endowment for International Peace publication, the Magic Kitten, the Iranian cyberespionage element, operates under the same digital cyber signature patterns and has the capabilities to inflict disruptive attacks. Russian, Chinese, and North Korean cyber experts are highly likely providing support to Iran.

c. According to the Wisconsin Project on Nuclear Arms Control, Iran's missiles and, in general, ICBM signatures detection, are likely to see only moderate to routine upgrades within the next ten years. Projected ICBM capabilities will likely resemble the same signatures as the old Soviet and North Korea Taepodong missile launching platforms and rocket capabilities.

d. **North Korea**. It is *highly likely* North Korea's top three impactful signatures will be associated with physical, visual, and speed. The top North Korean capabilities to be affected by these signatures will be their stockpile of missiles, encompassing both long- and short-range strike as well as stealth abilities, see figure 6. To view complete findings and analysis on North Korea, see page 107 and chart's nine and 10.

Estimated Signature and Capability Threat Matrix - North Korea							
	Short-Range Strike	Missiles	Long-Range Strike	Artificial Intelligence	Electronic Warfare	Unmanned Systems	Cyber
Physical	4	4	4	3	3	3	3
Visual	4	4	4	3	3	3	3
Speed	4	4	4	3	3	3	3
Electronic (Digital/Software)	3	3	3	3	3	3	3
Radio waves	3	3	3	3	3	3	3
Frequency	2	3	2	3	3	3	3
Digital	3	3	3	2	2	2	3
Thermal	3	4	3	2	2	2	3
Wavelength	3	2	2	3	3	3	3
Acoustic	3	3	3	2	2	2	3
Bio/Chemical	3	3	3	2	2	2	3
Infrared	3	3	3	2	2	2	3
Laser	3	3	3	2	2	2	3
Doctrine and other Non-Technical Process	3	3	3	2	2	2	3
Radiation	3	3	3	2	2	2	3
Seismic	3	3	3	2	2	2	3
Smell	3	3	3	2	2	2	3
Metanalytic	3	3	3	2	2	2	3
Ultra Violet	3	3	3	2	2	2	3
Hyperspectral	3	2	3	2	2	2	3
Magnetic	3	2	3	2	2	2	3
x-Ray	3	2	3	2	2	2	3

Figure 6 North Korea's Top Signature and Capability Threats in the next 10 to 20 years

5 Much More Difficult to Detect	4 More Difficult to Detect	3 Moderate to Routine Improvement	2 Easier to Detect	1 Extremely Easier to Detect

i) Based on Brookings published reports, by 2040, a greater range, speed, and a higher heat signature associated with a larger rocket engine are likely.

ii) According to US State Department estimates, while it remains among the poorest countries in the world, North Korea spends nearly a quarter of its gross domestic product (GDP) on its military.

iii) According to US Treasury reporting, the use of black-market procurement and their cyber hijacking of currency, which ranges in the billions per year, support their missile program advancements.

e. **Other Global Threats**. Improvements in cyber, AI, and other technologies will likely provide opportunities for weapons proliferation and great power competition in the coming years. GAE, in addition to China and Russia, are continuously leveraging arms sales and domestic technology to enhance their economy and alter both regional and world order.

i) France, Germany, US, Russia, and China were rated in the top five weapons exporters from 2014 to 2018, while some of the top importers included South Korea, Israel, India, and Saudi Arabia. France, Germany, and Spain have been working together to develop Air Combat Cloud as part of Future Combat Air System, which will, in real-time, connect and synchronize all the platforms and enable the processing and distribution of information to enhance situational awareness and collaborative operations. Additionally, according to Pierre Tran of Tech Watch and a report with Defense News, France is said to have increased investment in AI for future weapon systems.

ii) Additive manufacturing, also known as 3D printing, AI, and robots are driving the development of and use in a wide variety of weapons that are likely to have a significant impact on future threat capabilities.

iii) The significant evolution and advancement in three critical technologies, artificial intelligence, stealth technology, and emerging deception practices, are likely driving the bulk of the changes in the signatures that are becoming harder to detect.

What sensors and systems will the US Army likely need in order to detect, recognize, analyze, and target future threat signatures

1. The US Army will need a combination of multi-functional sensors and systems that are networked and integrated into the right places (air, land, space) connecting processed information at the point of data collection to recognize, detect, analyze and target the seven highly likely future threat capabilities (EW, Cyber, Missile, Unmanned, Deception, AI, and C4ISR programs) across the highly likely four future difficult threat signatures (physical, visual, electronic and other non-technical) in 2030-2040. Figure 1 on page 1 depicts the estimated overall global threat countries and signature capabilities. These highest likely difficult to detect capabilities and signatures account for 32% of all signatures and capabilities examined and recommend future sensors and systems to focus on these associated signatures and capabilities.
   a. According to one of the leading electronic surveillance companies, Era, it is highly likely to analyze and target future EW threat signatures, sensors, and systems technology must be able to collect multiple radio frequency sources across a significant bandwidth range, with multiple remote stations and a central processing station. These signals can be radars, data links, as well as identification friend or foe transponders, meaning even stealth aircraft are likely visible.

b. Samaras, and other leading experts from the Molecular Diversity Preservation International (MDPI) affirmed that detecting missile attacks signatures will likely be more effective through the employment of multiple integrator sensors. Figure 7, from a c-UAV analysis, demonstrates the capabilities of each sensor and system when employed independently.

c. Advanced data processing technologies with sensors and systems will enable

Characteristics

Detection Methods	Range	Position accuracy	Classification	Autonomous targets	Multiple targets	Low visibility conditions	Price
Human surveillance	**	***	*****	✓	✗	✗	****
Passive Electro-optical/infrared	***	****	****	✓	✗	✗	*
Acoustic	*	**	**	✓	✓	✓	***
Active Radar	****	****	***	✓	✓	✓	**

Figure 7 Comparison Analysis of Individual Sensors (Deep Learning of Multi-Sensor Data)

surveillance coverage rates necessary to confront future threats. Due to wide-band communication capabilities via satellite or terrestrial channels that provide surveillance products on-demand to units in the field, sensors, and systems in 2030-2040 will require the data storage and the tools necessary to exploit automated targeting.

2. **Sensors of the Future.** Rapid multi-sensor data fusion that combines information from multiple sensors and sources to achieve information and target dominance through algorithms and a variety of sources (HUMINT, CI, IMINT, GEOINT, MASINT, SIGINT, OSINT, TECHNINT, CYBINT/DNINT, and FININT) is paramount. Recognizing the development of organic sensor platforms to support ISR in radar, electro-optics, acoustics, and SIGINT enabled capability by computer and communications technologies will likely remain critical in 2030-2040. According to Wiley Online Library and 206 academic researchers in published sensor research, significant progress has been made in the development and advancement of flexible and multifunctional sensors that can achieve high sensitivity, accuracy, mechanical flexibility, and low cost. Based on reports published by Brookings and Talal Husseini from army technology.com to avoid detection from high-end thermal imaging and MASINT, threat countries have placed increasing importance on developing threat signature absorbing, camouflage and deception materials to avoid sensor detection.

   a. Based on these findings specifically, for the US Army to detect, recognize, analyze, and target future threat signatures in 2030-2040, sensors, and systems that are integrated and connected such as:

      i. Bistatic radar (hybrid systems involving space-borne radar illuminators and stealthy UAVs carrying bistatic receivers and signal processors); Infrared thermography (IRT) such as thermal imaging, and thermal video are examples of infrared imaging science; Quantitative Imaging Technology; Remote and Change Detection Sensing; EO / IR Sensors; MASINT Sensors; and SAR systems.

ii. It is likely radar and electro-optical systems will be essential for future reconnaissance and surveillance platforms to detect future threats.

iii. Due to increased capability from our adversaries and reduced armored signatures, adversaries are significantly trying to decrease its visual, thermal, electronic, digital, and radio wave threat emissions and signatures to conceal its speed, composition, and intentions. Therefore, sensors and systems must likely account for the actual physical, visual, thermal, and electronic changes as illustrated in Figure 7.

3. **Systems of the Future.** As previously discussed, it is *highly likely* there will not be a single system to discern all future signatures needed to identify or thwart enemy capabilities. While sensor systems will individually have weak signals, the best avenue to ensure correct identifiability or action is to integrate multiple weak signals from multiple sensors into a connected and integrated system to achieve precision. In 2030-2040 the automatic fusion of real-time sensor and intelligence data using basic sensor technology or exploitation technology via systems of systems will be of importance to mitigate the likely future threats.
    a. It is likely that in 2030-2040 the US and our adversaries will become increasingly dependent on technological force multipliers due to the numbers of personnel and equipment declining in response to economic pressures. With capabilities available in the open commercial marketplace, surveillance and reconnaissance are likely two critical capabilities that will undergo dramatic growth in performance as a result of the explosion in information technology.
    b. Algorithm designs and signal processing software will be of increasing focus for sensors and systems.
    c. Radar technology development is likely to continue its evolutionary pace in 2030-2040 and will likely become increasingly used. It is likely for any radar in the future, the radar signal processor will be the most critical element.

4. Due to sensor resolution improvements and as processing technology continues to advance, enabling more sophisticated algorithms will likely provide a critical capability for SAR sensors and systems to use automatic change detection in 2030-2040. The signal processors necessary to generate imagery and the data links necessary to disseminate it will be a critical factor in sensors and systems to achieve targeting identification and automating timely detection and targeting.

Table of Contents

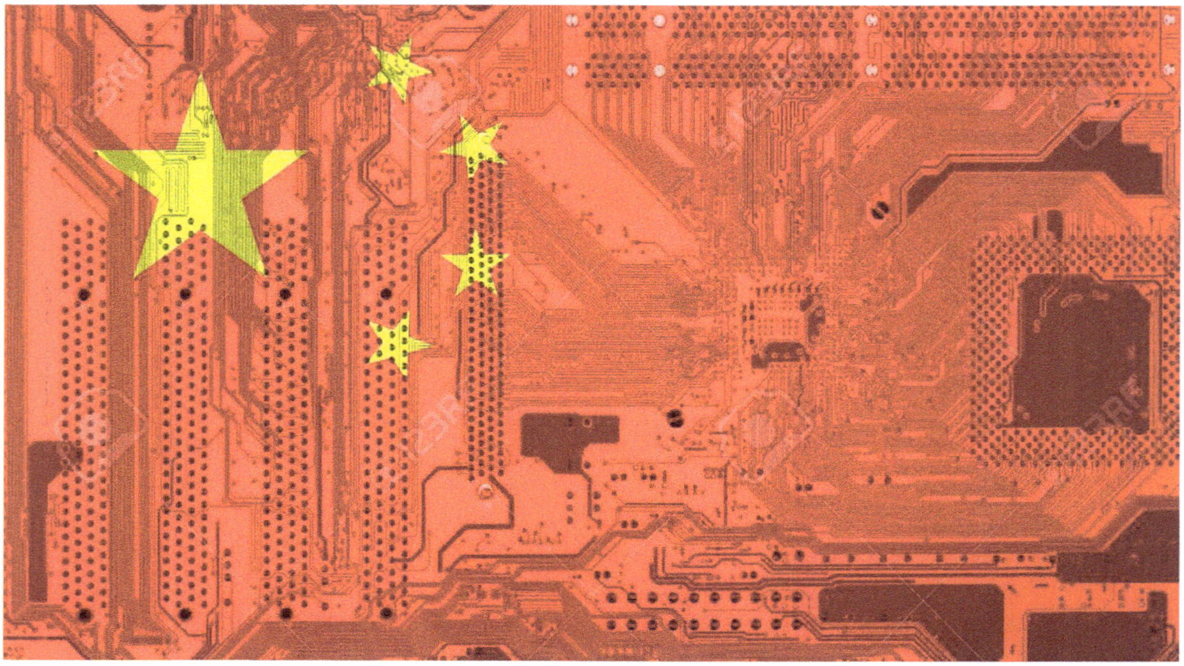

CHINA

China's Estimative Key Findings

Authored By: Russell Hoff

Based on an analysis of China's stated modernization guidance, defense spending priorities, and efforts pursuant to stated national goals in an evaluation of 19 capabilities against 22 possible signatures, it is highly likely that top four (4) most impactful signatures associated with Electronic (digital/software), Frequency, Radio Wave, and Doctrine & Non-technical Process signatures will both impact the most capabilities and become difficult, on average to detect in the 2030-2040 timeframe. See Figure 1 below

In fact, 15% (62/418) of future Chinese threat signatures, distributed over 12 of the 19 threat capabilities, will likely become difficult to detect between 2030 and 2040. It

	C4ISR	Electronic Warfare	Hypersonic	Stealth	Autonomous	Anti-Satellite	Unmanned Systems	Artificial Intelligence	Cyber	Quantum	Deception	Air Defense	Long-Range Strike	Missiles	Short-Range Strike	3D Printing	Batteries	Chemical	Tanks
ctronic (Digital/Software)	4	4	3	3	3	3	3	4	4	4	3	4	3	3	3	3	3	3	3
quency	4	4	4	4	4	3	4	3	3	3	3	3	3	3	3	3	3	3	3
lio waves	4	4	4	4	4	3	4	3	3	3	3	3	3	3	3	3	3	3	3
trine & Non-Technical Process	4	2	4	4	3	4	3	3	3	4	4	4	3	3	3	3	3	3	3
ital	4	4	3	3	3	3	3	4	4	4	3	3	3	3	3	3	3	3	3
sical	4	4	4	3	3	4	4	3	3	3	4	2	3	3	3	3	3	3	3
ual	4	4	3	3	4	3	4	3	3	3	4	2	3	3	3	3	3	3	3
velength	4	4	3	4	4	3	3	3	3	3	3	3	3	3	3	3	3	3	3
liation	4	4	3	3	3	3	3	3	3	3	3	4	3	3	3	3	3	3	3
tanalytic	3	4	3	3	3	3	3	4	4	3	3	3	3	3	3	3	3	3	3
ed	4	3	3	3	4	3	3	3	3	3	3	3	3	3	3	3	3	3	3
erspectral	4	3	3	3	3	3	3	3	3	3	3	4	3	3	3	3	3	3	3
rmal	2	2	4	4	3	4	3	3	3	3	3	3	3	3	3	3	3	3	3
er	3	3	3	3	3	4	3	3	3	3	3	3	3	3	3	3	3	3	3
ared	2	2	4	4	3	4	3	3	3	3	3	3	3	3	3	3	3	3	3
ustic	3	3	3	3	3	3	3	3	3	3	3	3	3	3	3	3	3	3	3
/Chemical	3	3	3	3	3	3	3	3	3	3	3	3	3	3	3	3	3	3	3
gnetic	3	3	3	3	3	3	3	3	3	3	3	3	3	3	3	3	3	3	3
smic	3	3	3	3	3	3	3	3	3	3	3	3	3	3	3	3	3	3	3
ell	3	3	3	3	3	3	3	3	3	3	3	3	3	3	3	3	3	3	3
ra Violet	3	3	3	3	3	3	3	3	3	3	3	3	3	3	3	3	3	3	3
ay	3	3	3	3	3	3	3	3	3	3	3	3	3	3	3	3	3	3	3

5 Much More Difficult to Detect	4 More Difficult to Detect	3 Moderate to Routine Improvement	2 Easier to Detect	1 Extremely Easier to Detect

Figure 1 Chinese Detection Estimate of Threat Capability's Signatures

should be noted that it is unlikely that any of the 418 signatures of the 19 capabilities will be on the extremities of the Future Signature Detection Scale, that is, to be VERY difficult to detect or VERY easy to detect between 2030-2040.

1. Driving these changes are several key facts. Specifically:

a. China's national leadership is prioritizing expenditures on key civil-military modernization efforts towards goal of "national rejuvenation" as stated in published national documents. President Xi Jinping, in his address at the 19th National Congress of the Communist Party of China, stated, "We must place greater focus on combat, encourage innovation, build systems, increase efficacy and efficiency, and further military-civilian integration."
    i. According to Nasdaq News, China is the second-largest economy in the world and the fastest-growing trillion-dollar economy. With a GDP of $14.14 trillion in 2019, it makes up 16.38% of the global economy. When compared on the basis of purchasing power parity (PPP), China is the largest economy with a GDP (PPP) of $27.31 trillion. Based on 2019 figures, the size of China's nominal GDP was lesser than that of U.S. by around $7.3 trillion, the gap is expected reduce to around $4.5 trillion by 2024. Nasdaq News further assesses that China is on the path to become a $20 trillion economy by 2024.
    ii. China's July 2019 National Defense in the New Era highlighted a strategic goal of "Efforts will be made to advance the integrated development of mechanization and informationization, speed up the development of intelligent military, create a modernized military force structure with Chinese characteristics, improve and develop socialist military institutions with Chinese features, and constantly enhance the capabilities to fulfill the missions and tasks in the new era."
b. China's close control over national academia and commercial sector enables focused R&D and modernization efforts and fusion of civil-military, dual use technologies.
    i. The International Institute for Strategic Studies group assess that China is investing heavily in its pursuit and integration of emerging dual-use technologies, hoping they will help the People's Liberation Army (PLA) to surpass conventional military capabilities to achieve battlefield dominance across domains. Further, technologies such as AI, cyber infrastructure and software, and automation are primarily civilian in their application, but their relevance to defense and to how future wars will be fought is clearly growing.
c. China's theft of intellectual property, reverse-engineering, and incorporation of outside (US & other) academic & scientific training accelerates modernization efforts.
    i. Defense Secretary Mark Esper recently warned that China was perpetrating the "greatest intellectual property theft in human history."
d. Despite many broad-ranging efforts to modernize its military and potential overstatements of true capabilities, China is making headway and is becoming more of a challenge, which includes the integration of enabling technologies.
    i. At the December 31, 2015 inauguration ceremony in Bejing, President Xi Jinping pointed out that the establishment of the PLA Army's

leading organ, the PLA Rocket Force and the PLA Strategic Support Force was an important decision made by the CPC Central Committee and the CMC to realize the Chinese Dream and the Dream of a Strong Military, and a strategic initiative to build a modern military power system with Chinese characteristics. The Institute for National Strategic Studies (INSS) assesses the Strategic Support Force (SSF) role is to integrate the PLA's space, cyber, and EW domains, to enabling multi-domain war-fighting capabilities, to fight and win future informatized wars.

2. China's C4ISR is likely to see the most upgrades, and consequently, the most negative overall change in the capability's signature profile, hence more difficult to detect. In fact, in 11 of the 22 signatures examined, China's C4ISR capability is likely to get more difficult to detect over the next 20 years. These upgrades will likely impact not only the four (4) most impactful signatures named above but also a wide variety of other signatures including those such as non-technical signatures such as Digital, Physical Visual, Wavelength, Radiation, Speed, and Hyperspectral. These changes are largely due to forecasted improvements in enabling technologies and capabilities such as Cyber and Artificial Intelligence.
    a. China created the Strategic Support Force (SSF) in December 2015 to integrate the PLA's space, cyber, and electronic warfare domains. Both the Institute for National Strategic Studies (INSS) and analysts for the Cyber Defense Review assess that the Network System Department of the SSF is likely the convergence of PLA's capabilities for cyber, electronic, and psychological warfare into a single force which likely enable it to take advantage of synergies among operations in these domains.
    b. Due to the political system of the Communist Party of China there is likely a high degree of collaboration and control in China among academia, the private sector and government and military organizations.
    c. According to the 2019 China's State Council Information Office issued national defense white paper "China's National Defense in the New Era," China's national focus is to apply cutting edge technologies such as artificial intelligence, quantum information, big data (metadata), cloud computing, and the Internet of Things into the military field (i.e. weaponization of enabling technologies).
    d. Jane's 2020 Sentinel Security Assessment - China And Northeast Asia assess that integrated joint exercises are being held throughout China in which select units from all the services, as well as the reserve forces, are making use of significantly improved IT capabilities to practice true joint operations. The same report notes EW and cyber warfare are becoming a more common feature in PLA exercises to replicate complex electromagnetic spectrum conditions.

e. Dean Cheng, the Senior Research Fellow in the Asian Studies Center Davis Institute for National Security and Foreign Policy at the Heritage Foundation testified before U.S.–China Economic and Security Review Commission that, "It is [integrated electronic warfare (INEW)] methods that use a combination of electronic warfare and network warfare techniques to attrit and disrupt the adversary's networked information systems, while defending one's own, in order to secure information dominance over the battlefield."

3. Analyst applied a cursory analysis of signature changes of 3D printing, Batteries, Chemical, Long- and Short-range strike, Missiles, and Tanks capabilities opting for a focused emphasis in analyzing modernization of capabilities that China is allocating more funding towards. Analyst has low analytical confidence in signature change ratings for 3D printing, Batteries, Chemical, Long- and Short-range strike, Missiles, and Tanks.

4. When a capability category whose aggregate signature average falls below a composite average of all signatures, the category is estimated as inconsequential when determining overall signature detection difficulty. For China, Air Defense capability falls below the composite average of 3.13. When Air Defense and low confidence capabilities are removed from calculations an interesting finding emerges: the percentage of forecasted, harder to detect signatures increases to 24% (58/242), up from 15% (62/418). See figure 2 below:

5. China's C4ISR Capabilities Likely to Make Signature Detection More Difficult in 11 of 22 Analyzed Signature Categories by 2030; IR & Thermal Likely Easier to Detect.
    a. Due to the increasing convergence of artificial intelligence (AI), cyber, autonomous and unmanned systems, electronic warfare (EW), and space-based systems within China's military modernization efforts, China likely continues to gain C4ISR experience and capabilities augmented by equipment modernization efforts and practiced implementation of full-spectrum, integrated, multi-domain operations.

6. By 2030, China highly likely to improve EW capabilities; integrated EW tactics, enhanced by AI and cyber, likely makes detection in EM spectrum more difficult; IR and thermal signatures likely easier to detect; possible increased unit footprint signature.
    a. Due to China's focus on military modernization efforts, China is likely to offset its EW detection vulnerabilities with full-spectrum, integrated network electronic warfare (INEW) tactics, including the weaponized use of AI and cyber, but will likely still remain vulnerable in IR and thermal detection.

	C4ISR	Electronic Warfare	Hypersonic	Stealth	Autonomous	Anti-Satellite	Unmanned Systems	Artificial Intelligence	Cyber	Quantum	Deception	Air Defense	Long-Range Strike	Missiles	Short-Range Strike	3D Printing	Batteries	Chemical	Tanks
Electronic (Digital/Software)	4	4	3	3	3	3	3	4	4	4	3	4	3	3	3	3	3	3	3
Frequency	4	4	4	4	4	3	4	3	3	3	3	3	3	3	3	3	3	3	3
Radio waves	4	4	4	4	4	3	4	3	3	3	3	3	3	3	3	3	3	3	3
Doctrine & Non-Technical Process	4	2	4	4	3	4	3	3	3	4	4	4	3	3	3	3	3	3	3
Digital	4	4	3	3	3	3	3	4	4	4	3	3	3	3	3	3	3	3	3
Physical	4	4	4	3	3	4	4	3	3	3	4	2	3	3	3	3	3	3	3
Visual	4	4	3	3	4	3	4	3	3	3	4	2	3	3	3	3	3	3	3
Wavelength	4	4	3	4	4	3	3	3	3	3	3	3	3	3	3	3	3	3	3
Radiation	4	4	3	3	3	3	3	3	3	3	3	4	3	3	3	3	3	3	3
Metanalytic	3	4	3	3	3	3	3	4	4	3	3	3	3	3	3	3	3	3	3
Speed	4	3	3	3	4	3	3	3	3	3	3	3	3	3	3	3	3	3	3
Hyperspectral	4	3	3	3	3	3	3	3	3	3	3	4	3	3	3	3	3	3	3
Thermal	2	2	4	4	3	4	3	3	3	3	3	3	3	3	3	3	3	3	3
Laser	3	3	3	3	3	4	3	3	3	3	3	3	3	3	3	3	3	3	3
Infrared	2	2	4	4	3	4	3	3	3	3	3	3	3	3	3	3	3	3	3
Acoustic	3	3	3	3	3	3	3	3	3	3	3	3	3	3	3	3	3	3	3
Bio/Chemical	3	3	3	3	3	3	3	3	3	3	3	3	3	3	3	3	3	3	3
Magnetic	3	3	3	3	3	3	3	3	3	3	3	3	3	3	3	3	3	3	3
Seismic	3	3	3	3	3	3	3	3	3	3	3	3	3	3	3	3	3	3	3
Smell	3	3	3	3	3	3	3	3	3	3	3	3	3	3	3	3	3	3	3
Ultra Violet	3	3	3	3	3	3	3	3	3	3	3	3	3	3	3	3	3	3	3
X-Ray	3	3	3	3	3	3	3	3	3	3	3	3	3	3	3	3	3	3	

5 Much More Difficult to Detect	4 More Difficult to Detect	3 Moderate to Routine Improvement	2 Easier to Detect	1 Extremely Easier to Detect

Figure 2 Chinese Detection Estimates of Threat Capability's Signatures Adjusted Based on Categorical Averages

7. China's hypersonic weapon advancements throughout 2030-2040 are likely to make radio waves, thermal, IR, and non-technical signatures more difficult to detect.
    a. Sustained budget increases and rapid technical improvement of a relatively new technology likely enables China's hypersonic weapon advancements to progress, albeit relatively slowly.

8. China likely to achieve enhanced stealth and camouflage technology by 2030.
    a. Chinese President Xi Jinping demanded the complete modernization of the Peoples Liberation Army by 2035 thereby emphasizing needs for advanced materials and alternative energy such as stealth and camouflage materials.

9. China's modernization of autonomous capabilities will make frequency, radio wave, visual, and wavelength signatures harder to detect by 2030.
    a. Due to their the most aggressive and advanced autonomous military program to date, China likely is increasing its domestic development abilities of autonomous systems through innovation incentives and technology theft.

10. China's mobile, ground-based, anti-satellite, laser weapon capability likely to make thermal, IR, laser, physical, non-technical signatures harder to detect within 15 years.

   a. Despite directed energy weapons being a lower, national-level priority than other military projects, China likely continues its technological advancements in ground based, anti-satellite technology efforts.

11. China's frequency, radio, physical, and visual signatures of Unmanned Area Vehicles (UAVs) will likely become difficult to detect within the next 15 years
    a. China has opened UAVs facilities to reverse engineer and improve technologies to build their own technology capacity, which likely translates to improved UAV capabilities and the capacity for mass production.

12. China is likely to reduce electronic, digital, and metanalytic signatures by weaponizing artificial intelligence by 2030.
    a. With China's close ties between its government and private sector firms enable China's defense and intelligence industries to focus AI research and implementation, using China's vast collection of data, China is likely to weaponize AI to aid military with intelligence analysis, decision-making, vehicle autonomy, weaponry, and information operations.

13. China is highly likely to reduce offensive cyber signatures by 2030 making detection and attribution highly likely to be more difficult.
    a. China's military modernization efforts towards meeting its stated, national goals, is likely to continue to see improvements in offensive cyber capabilities, cyber espionage, and cyber-attacks.

14. The U.S. Army will likely need Measurement and Signature Intelligence (MASINT) sensors and systems that can detect, track, identify or describe the distinctive characteristics of Chinese fixed or dynamic Electronic (digital/software), Frequency, Radio Wave, and Doctrine & Non-technical Process signatures.

15. It is likely that Infrared Signatures of C4ISR advancements will be easier to detect. It is likely that Doctrine & Other Non-technical process, thermal, and IR signatures of electronic warfare advancements will be easier to detect. It is likely that China's physical and visual signatures of air defense advancements will be easier to detect.

China's C4ISR Capabilities Likely to Make Signature Detection More Difficult in 11 of 22 Analyzed Signature Categories by 2030; IR & Thermal Likely Easier to Detect

Executive Summary:

Due to the increasing convergence of artificial intelligence (AI), cyber, autonomous and unmanned systems, electronic warfare (EW), and space-based systems within China's military modernization efforts, China's C4ISR capabilities are likely to make signature detection more difficult across 11 of 22 analyzed signature categories by 2030. Signatures made more difficult to detect include much of the electromagnetic spectrum, digital, and doctrine & non-tactical processes. IR and Thermal signatures are likely easier to detect. Despite a relatively new (2015) reorganization of military forces and capabilities, China continues to gain C4ISR experience and capabilities augmented by equipment modernization efforts and practiced implementation of full-spectrum, integrated, multi-domain operations.

Discussion:

By 2030, China's focus on military modernization efforts will likely advance C4ISR capabilities which includes integrating enabling technologies such as artificial intelligence (AI), cyber, autonomous and unmanned systems, electronic warfare (EW) systems, and space-based architecture.

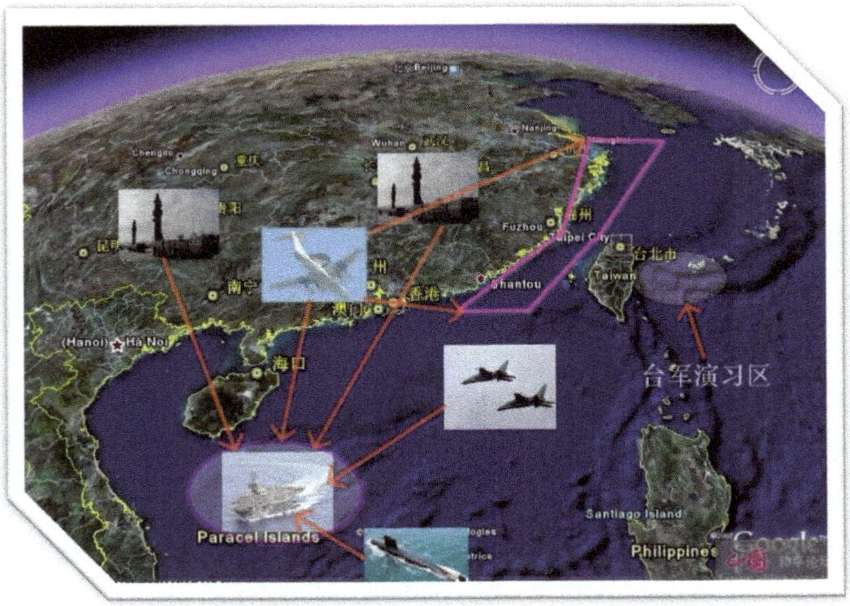

Figure 1: From a Chinese source discussing the benefits of the SSF and integration of capabilities. Source: Newton.com

China created the Strategic Support Force (SSF) in December 2015 to integrate the space, cyber, and EW domains to enable multi-domain war-fighting capabilities. [H] The unit's mission reportedly is seizing and maintaining battlefield information control. [H] The SSF's ability to provide the information umbrella of space-based C4ISR, intelligence support, and battlefield environment assessments helps forge a common intelligence picture among joint forces within each theater command (see Figure

3), a fundamental requirement for fulfilling the People's Liberation Army's (PLA) mission of winning "informationized local wars." H

Figure 4: China displays various radar technologies, some claimed as anti-stealth radars. Source: globalsecurity.org

Despite the relatively new reorganization of military forces and capabilities, China continues to gain C4ISR experience and capabilities augmented by equipment modernization efforts and practiced implementation of full-spectrum, integrated, multi-domain operations. EW and cyber warfare are a more common feature in PLA exercises to replicate complex electromagnetic spectrum conditions.[1] See China EW article, See China cyber warfare report for additional information. Chinese PLA are modernizing special forces with the next-generation technology available to conduct the full spectrum of future special operations.[2] The PLA likely gains synergy from full-spectrum, integrated network electronic warfare (INEW) tactics (see Figure 4), including use of unmanned systems (see China UAV report), autonomous systems (see China autonomous information), space-based reconnaissance or communication, EW systems including aircraft (see China EW report), and the weaponized use of AI (See China AI article) and cyber (China cyber insert). Synergy also provides benefits from the signature reduction each modernized capability provides (see individual capability statement). IR and thermal signatures are likely to be easier to detect, especially in EW equipment (see China EW article); however, it is likely that China will offset vulnerabilities with enabling technologies such as AI, cyber, stealth, deception practices, and, eventually, quantum capabilities.

---

[1] "Jane's Sentinel Security Assessment - China And Northeast Asia: China – Army – Military Exercises", in Jane's (accessed through USAWC Subscription Databases): "[Chinese] Cyber and EW forces are playing a more prominent role in exercises too. The PLA's first major wargames involving cyber forces were held at Zurihe in Inner Mongolia in June 2013, and they included SOF, army aviation, and EW and digital units. EW and cyber warfare are becoming a more common feature in PLA exercises to replicate complex electromagnetic spectrum conditions." Source Reliability: High.

[2] "Jane's Amphibious and Special Forces: Capability Analysis", in Jane's (accessed through USAWC Subscription Databases): "The emergence of the SSF provides an interesting glimpse into the future roadmap of the Chinese SOF. Acknowledgement of the future strategic importance of cyber warfare and EW in particular shows how Chinese SOF are seeking to attain all the next-generation technology available to conduct the full spectrum of future special operations. However, integration into the various Special Operations Brigades and its subsequent connectivity across not only SOF, but also conventional units, will decide upon the degree to which it is used across the contemporary and future operating environments as the PLA continues to seek influence beyond its borders." Source Reliability: High.

Analytic Confidence:

The analytic confidence for this estimate is *moderate*. Open sources were used and the reliability of the sources available on this topic were above average with several high-quality sources available for the estimate. Source content varied; however, sources tended to corroborate one another. There was adequate time allotted to the task and the analyst used a structured research method.

Authored by: Russell Hoff

China Highly Likely to Improve EW Capabilities; Integrated EW Tactics, Enhanced by AI and Cyber, Likely Makes Detection in EM Spectrum More Difficult; IR and Thermal Signatures Likely Easier to Detect; Possible Increased Unit Footprint Signature

Executive Summary:

Due to China's focus on military modernization efforts, China is highly likely to improve electronic warfare (EW) capabilities by 2030. Despite pursuing high-powered EW jamming equipment that likely increases detectable radio frequencies, thermal, and Infrared (IR) signatures of some platforms, China is likely to offset those vulnerabilities with full-spectrum, integrated network electronic warfare (INEW) tactics, including the weaponized use of AI and cyber. Overall, signatures made more difficult to detect include much of the electromagnetic spectrum.

Discussion:

By 2030, China's focus on military modernization efforts, across all domains, is highly likely to improve EW capabilities (Figure 5). Despite pursuing high-powered, ground-based EW systems (especially jammers) based on vacuum-tube technology instead of solid-state electronics which likely increases detectable RF, IR, and thermal signatures, China is likely to offset those vulnerabilities with full-spectrum, integrated network electronic warfare (INEW) tactics, including the weaponized use of AI and cyber (See China AI insert; See China cyber report).

Figure 5: PLA EW systems on display during China's military parade on 1 Oct 2019. Source: CSIS

Vacuum tubes function at higher frequencies and higher power levels than solid-state EW systems M,[3] which enables EW systems to emit jamming signals at enormous power

[3] "Analysis: Western forces vulnerable to China's pursuit of vacuum tube technology", Jane's Defence Weekly 12-Feb-2020 (accessed through USAWC Subscription Database): "China has invested heavily in vacuum tube technology because there are limitations to the

levels.[4] EW specialists explained that signals from vacuum tube-powered devices operate with higher precision at the millimeter wavelength and transmit at higher amplitude, making them more difficult to jam.[5] While use of vacuum tubes increases detectable RF power levels, it also increases detectable thermal and IR signatures from the expansive heat signatures M vacuum tubes produce over solid-state technology.

China created the Strategic Support Force (SSF) in December 2015 to integrate the PLA's space, cyber, and EW domains, to enabling multi-domain war-fighting capabilities, to fight and win future informatized wars. H The unit's mission reportedly is seizing and maintaining battlefield information control. H The Network System Department of the SSF likely is the convergence of PLA's capabilities for cyber, electronic, and psychological warfare into a single force which likely enable it to take advantage of key synergies among operations in these domains.[6] Merging the capabilities possibly increases a unit's signature in terms of personnel and equipment needed to gain synergies across the domains, especially towards controlling complex electromagnetic environments. H

China adopted and implemented the concept of INEW which combines network warfare and EW in a holistic, complementary approach towards information warfare dominance. H It is likely the PLA will leverage AI and cyber to gain and analyze adversarial information using machine learning and meta-analytics. A likely application of INEW is leveraging AI and machine learning in dynamic spectrum management to learn and rapidly devise countermeasures for adversary systems. H AI will be critical in cognitive radio application to optimize channel detection and signal selection in a complex and crowded electromagnetic spectrum. H Increased use across the electromagnetic spectrum

solid-state designs that exist today. "Vacuum tubes will function at higher frequencies and higher power levels than any solid-state EW system," [Malcolm Carruthers of the US microwave electronics firm Photonis]." Source Reliability: High.

[4] "Analysis: Western forces vulnerable to China's pursuit of vacuum tube technology", Jane's Defence Weekly 12-Feb-2020 (accessed through USAWC Subscription Databases): "Both Chinese and Russian EW systems can emit jamming signals at enormous power levels. This has been seen in the military operations by Russian units and their proxies since 2014 in Eastern Ukraine. Some ground-based Russian EW systems have been spotted in reconnaissance photographs in the Donbass region where they were parked next to a central utility compound so they could be plugged directly into the main electrical power grid of a major city. The ability of an EW system to use this much power is because of the vacuum tubes that can support transmissions at such amplitudes." Source Reliability: High.

[5] "Analysis: Western forces vulnerable to China's pursuit of vacuum tube technology", Jane's Defence Weekly 12-Feb-2020 (accessed through USAWC Subscription Database): "More than one EW specialist explained to ***Jane's*** that signals from vacuum tube-powered devices operate with higher precision at the millimetre wavelength and transmit at higher amplitude, making them more difficult to jam. This puts US and other allied armed forces in increasing jeopardy because if the West "is not embracing the vacuum electronic tube, you are not going to be able to face these threats", according to [Malcolm Carruthers of the US microwave electronics firm Photonis]. "In the West there is a stigma attached to this industry because it is looked upon as 'old technology' and yet the enemy is still expanding on it and using it."" Source Reliability: High.

[6] "The Strategic Support Force and the Future of Chinese Information Operations", Army Cyber Institute, The Cyber Defense Review, Vol. 3, No. 1 (SPRING 2018), pp. 105-122 (accessed through USAWC Subscription Database): "Concurrently, the SSF has integrated the PLA's capabilities for cyber, electronic, and psychological warfare into a single force within its Network Systems Department (网络系统部), which could enable it to take advantage of key synergies among operations in these domains. However, beyond the SSF, the PLA also appears to be building up network-electronic operations (网电作战) capabilities within its national Joint Staff Department headquarters and within new regional theater com-mands (战区), reflecting the emergence of a multi-level force structure specializing in information operations. Thus, the SSF reflects the PLA's uniquely integrated approach to force structure and operations in these vital new domains. This realization of this paradigm through the SSF will enhance the PLA's capabilities to fight and win future "informatized" (信息化) wars." Source Reliability: High.

likely presents a future detectable signature should future, US EW AI systems advance to detect and counter.

China demonstrated advancements in EW through airframe modernization efforts of the 2015 J-16D variant [M] and the 2017 Chengdu J-20[M] (Figure 6) potentially outfitted with electronic attack capabilities. China modified some H-6G bombers to an electronic warfare configuration fitted with electronic counter measure pods that can engage in combat missions using electronic jamming, suppression, and anti-radiation [measures]; China's JH-7 fighter-bomber has also been seen carrying such ECM pods in PLA Air Force exercises.[7] Additionally, China is pursuing heavy fuel engines for UAVs which potentially extends the air vehicle's endurance and payload capacity which have potential for EW mounted systems.[8]

Figure 6: China could develop their Chengdu J-20 Fifth Generation Fighter Jet into an electronic attack jet. Source: The Diplomat

According to the 2016 China Military Power Report, China's EW weapons include "jamming equipment against multiple communication and radar systems and GPS satellite systems. EW systems are also being deployed with other sea- and air-based platforms intended for both offensive and defensive operations." [H] Additionally, EW and cyber warfare are becoming a more common feature in PLA exercises to replicate

[7] "[China's People's Liberation Army Navy (PLAN)] deploys H-6G bomber in electronic warfare configuration", Jane's Defence Weekly 25-Jan-2018 (accessed through USAWC Subscription Database): "Equipped with an electronic countermeasure (ECM) pod underneath each of its wings, the aircraft was dispatched by the PLAN's South China Sea Fleet on exercises in December 2017, marking the first time that the bomber played a supporting electronic warfare role, the state-owned Global Times newspaper reported on 21 January, citing a China Central Television (CCTV) programme. "The modified H-6G fitted with ECM pods can engage in combat missions using electronic jamming, suppression, and anti-radiation [measures]," the paper quoted the CCTV programme as saying, adding that China's JH-7 fighter-bomber has also been seen carrying such ECM pods in PLA Air Force exercises." Source Reliability: High.

[8] "Keeping pace: UAV engines evolve to meet platform demands", Jane's International Defence Review 18-Mar-2020 (accessed through USAWC Subscription Database): "However, while available payloads have evolved and expanded rapidly from basic electro-optical/infrared (EO/IR) sights to encompass advanced mission systems such as electronic warfare (EW) equipment and synthetic aperture radar (SAR), propulsion technology has not typically progressed at a comparable rate. Instead, manufacturers have often been forced to rely on modified legacy aircraft engines that are not optimised for the stresses of extended operations performed by long-endurance UAVs....[China Academy of Aerospace Aerodynamics] is planning to install an indigenous [Heavy Fuel Engine (HFE)] propulsion system to further push the CH-4's flight envelope and performance. Although the company was unable to disclose specifics at this stage, Jane's understands that it is evaluating options for a 134–150hp class HFE that would extend the air vehicle's endurance to 50 hours and payload capacity to 400 kg, with a corresponding improvement in service ceiling to 29,527 ft." Source Reliability: High.

complex electromagnetic spectrum conditions.[9] Without US mitigation efforts, China's holistic EW approach likely complicates US ability to simultaneously counter EW across multiple domains.

Analytic Confidence:

The analytic confidence for this estimate is *moderate*. Open sources were used and the reliability of the sources available on this topic were above average with several high-quality sources available for the estimate. Source content varied; however, sources tended to corroborate one another. There was adequate time allotted to the task and the analyst used a structured research method.

Authored by: Russell Hoff

[9] "Jane's Sentinel Security Assessment - China And Northeast Asia: China – Army – Military Exercises", in Jane's (accessed through USAWC Subscription Database): "[Chinese] Cyber and EW forces are playing a more prominent role in exercises too. The PLA's first major wargames involving cyber forces were held at Zurihe in Inner Mongolia in June 2013, and they included SOF, army aviation, and EW and digital units. EW and cyber warfare are becoming a more common feature in PLA exercises to replicate complex electromagnetic spectrum conditions." Source Reliability: High.

China's Hypersonic Weapon Advancements Throughout 2030-2040 are Likely to Make Radio Waves, Thermal, IR, and Non-Technical Signatures More Difficult to Detect

Executive Summary:

Due to sustained budget increases and rapid technical improvement of a relatively new technology (2014), China's hypersonic weapon advancements throughout 2030-2040 are likely to make radio waves, thermal, IR, and non-technical signatures more difficult to detect. Despite to a plateau of major technological advancements, China's hypersonic weapon advancements are likely progressing, albeit slowly.

Discussion:

China is working on two classes of hypersonic weapons – hypersonic glide vehicles (HGV) and hypersonic cruise missiles (HCM). The DF-ZF (dubbed Wu-14 in the United States) is a HGV launched from an existing ballistic missile booster (such as the DF-17, figure 7) and then coasts unpowered using its own momentum at claimed speeds of Mach 10.[10]

Figure 7: Screen capture of DF-17 hypersonic ballistic missile during a parade in Beijing celebrating the 70th anniversary of the founding of the People's Republic of China on 1 Oct 2019. Source: [ll]

TheXingkong-2 (Starry Sky-2) is a HCM with a conventional booster rocket launched before separating and engaging its own engine (scramjet) to accelerate to claimed hypersonic speeds of up to Mach 6.[11]

---

[10] "Chinese hypersonic programme reflects regional priorities", Jane's Intelligence Review 12-Mar-2019 (accessed through USAWC Subscription Database): "The first known flight test of the WU-14 HGV (as the DF-ZF was identified in open sources at the time) occurred in 2014. A 28 December 2017 report on The Diplomat website cited an unnamed US government source who used the designation DF-17 to refer to the combined HGV and its medium-range ballistic missile (MRBM) booster. It cannot be confirmed that DF-17 is an official designation used by the Chinese military. …Although the basic function of the DF-ZF is similar to the US Hypersonic Technology Vehicle (HTV) or Advanced Hypersonic Weapon (AHW) and the Russian Avangard (Object [Project] 4202/Yu-71/15Yu71), there appear to be differences in their intended uses. Compared with the Russian and US systems – which would glide at speeds of about Mach 20 – the material available in open sources on the DF-ZF suggests that it would travel at about Mach 10."

[11] "China claims successful test of hypersonic waverider", Jane's Defence Weekly 10-Aug-2018 (accessed through USAWC Subscription Database): "[China Academy of Aerospace Aerodynamics (CAAA)] claimed that the test vehicle – which it said had been in development for three years – was launched from an undisclosed launch facility in northwestern China and separated cleanly from the booster rocket following a 10-minute ascent and controlled transition, successfully engaging its own propulsion system to perform independent flight for over 400 seconds, attaining a maximum speed of Mach 6 and flight ceiling of 30 km (98,425 km/h).

It is likely that future HGV ballistic missile booster and HCM launches will have similar acoustic, electrical, physical, infrared, thermal, and visual signatures as current ones; the main purpose of those devices are to get the hypersonic weapon to altitude before accelerating to ultra-high speeds. H The detection of several signatures of the HGV likely will become more difficult.

A hypersonic weapon's speed creates several signatures. Speed is a signature which will remain routinely detectable. Speeds greater than Mach 5 creates frictional heating of the weapon by the atmosphere. Hypersonic weapons create large IR and thermal signatures due to frictional skin heating, which would make them more easily detectable by thermal and IR sensors;M however, to mitigate excessive temperatures of the hypersonic weapons, very special alloys, such as nickel or composites, or active cooling, by circulating fuel through the skin, is used to absorb the heat. H Hypersonic travel is so brutal that an object traveling at such speeds slowly tears itself apart during flight as the speed magnifies heat, wind, and other environmental factors. This gradually alters a hypersonic weapon's flight dynamics, making accuracy an increasingly difficult problem. H At hypersonic speeds, the exterior temperature of a hypersonic vehicle or weapon can exceed 2,000°F, necessitating advanced materials that will protect interior electronics. H Chinese scientists have developed a new heat-resistant material for hypersonic aircraft that can endure more than 3,000°C from heating caused by its passage through the atmosphere at speeds between Mach 5 and Mach 20.[12] These special materials will make detection of thermal and IR signatures more difficult in the future. Distance covered by hypersonic speeds requires detection far enough out for defensive measures. A byproduct of speed is the signature of shock waves of air density changes and the distortion of the shock waves based on attack angles. M

Another byproduct of speed is plasma. At high hypersonic speeds (greater than Mach 8 to 15), H the molecules break apart producing an electrically charged plasma around the aircraft. H Plasma affects radar cross-section likely making the hypersonic weapon more difficult to detect. The plasma has various electromagnetic resonance frequencies depending on the temperature and density of the gas and the thickness of the plasma layer to absorb matching radio frequencies, reflecting off if frequencies are higher or lower. M

A hypersonic weapon needs a guidance signal to maneuver to the intended target. From the speed and high altitude, it is likely the guidance signal transmits through satellites

[12] "China claims development of new heat-resistant material for hypersonic aircraft", Jane's Defence Weekly 29-Apr-2019 (accessed through USAWC Subscription Database): "Chinese scientists have developed a new heat-resistant material for hypersonic aircraft that can endure more than 3,000°C from heating caused by its passage through the atmosphere at speeds between Mach 5 and Mach 20, state-owned newspaper Global Times reported on 28 April. Fan Jinglian, the lead scientist in the project, was quoted by the paper as saying that the new material "outperforms all similar foreign-made ones with its high melting point, low density, and high malleability". The material, which is reportedly a composite of ceramics and refractory metals, "enables a hypersonic aircraft to fly at Mach 5–20 within the atmosphere for several hours, as the high heat resulting from the friction between the aircraft and the air reaches between 2,000°C to 3,000°C: a temperature normal metal would not be able to endure", stated the report.
"The combination of ceramics and refractory metals makes the material far more efficient than foreign-made ones, and this technology is world-leading," Fan, who is a professor at the Central South University in Changsha, Hunan Province, was quoted as saying."
Source Reliability: High.

from a ground station. The demand for reliable, continuous communications, through a plasma sheath created by ultra-high speeds has a wide market appeal for commercial space launch vehicles, defense weaponry, and commercial and defense aircraft. [H] The Missile Defense Agency (MDA) looked for an innovative solution to mitigate or minimize the effects of the plasma sheath on communications to and from a flight vehicle during in hypersonic flight – often leading to full communication blackout for standard radio frequencies used in telemetry and other communications. [H] The properties of the plasma sheath can vary during flight, complicating any efforts to generate resonance, but the researchers suggested the matched layer can compensate for these changes if it is made from a material whose electromagnetic properties can be altered electrically. [M] Additionally, autonomous hypersonic weapons imbedded with AI will likely reduce detectable radio signals emissions since commands would likely be self-contained within the platform, see China autonomous insert, see China AI article.

US hypersonic weapons have a length between just five and 10 feet, weighing about 500 pounds and encased in materials like ceramic and carbon fiber composites or nickel-chromium superalloys. [H] Using US hypersonic weapon sizes as a comparison, plus future size reduction technologies, it is likely the US will need a small radar cross section and more sensitive radars to detect similar or smaller Chinese hypersonic weapons. [M]

The People's Liberation Army (PLA) can ground launch hypersonic weapons from static sites as they have done during test launches[H] and likely, in the future, from mobile, transporter-erector-launchers (TEL).[M] Although the HGV paraded on TELs like in figure 7, it is unconfirmed if China conducted HGVs launches from a TEL. The TELs will likely be similar to those used by the PLA's Rocket Force (PLARF) since the hypersonic weapons, like conventional and nuclear launched weapons, likely will collocate with their formations. [M] The future mobility and the niche weapon intermingled with conventional weapons likely make detection of hypersonic weapons at the PLARF unit level more difficult.

Hypersonic missiles present a high military significant threat. Advances in hypersonic technologies have significant implications for national security. [H] The MDA has been directed to develop and deploy defenses against hypersonic and cruise missile threats as soon as technologically able. In 2016, the Air Force Studies Board, part of the U.S. National Academy of Science, published a report on the threat posed to U.S. interests by high-speed maneuvering weapons systems. [H] The report highlighted that because hypersonic weapons operate below the normal trajectories of ballistic missiles and above the speeds of non-hypersonic cruise missiles which exceed the capabilities of the United States' current missile and air defense systems and infrastructure; hypersonic weapons combine speed and maneuverability between the air and space regimes to produce significant new offensive capability that could pose a complex defensive challenge. [H]

Analytic Confidence:

The analytic confidence for this estimate is *moderate*. Open sources were used and the reliability of the sources available on this topic were above average with several high-quality sources available for the estimate. Source content varied; however, sources tended to corroborate one another. There was adequate time allotted to the task and the analyst used a structured research method.

Authored by: Russell Hoff

China Likely to Achieve Enhanced Stealth and Camouflage Technology by 2035

Executive Summary:

Due to a Chinese leadership focus on military modernization as well as a number of promising, ongoing initiatives, China is likely to achieve enhanced stealth and camouflage technology to defeat United States (US) military advancements by 2035. Chinese President Xi Jinping demanded the complete modernization of the Peoples Liberation Army by 2035 thereby focusing on advanced materials and alternative energy such as stealth and camouflage materials. Despite China slightly trailing the US on technological investments, China's close ties between its government and private sector firms likely enable China's defense and intelligence industries to focus camouflage research and implementation using China's vast collection of data.

Discussion:

Chinese scientists claiming stealth breakthroughs is a recent example of China's efforts to modernize its various warfighting platforms and defenses, as well as answering Xi Jinping's national direction. According to the South China Morning Post (SCMP), "Chinese scientists hail incredible stealth breakthrough that may blind military radar systems." [M] Currently, stealth objects rely on the geometric shape of its housing or specialized paint that can absorb radar waves. The composition, material, or housing deflects radar signals possibly affecting aerodynamics and agility, while paint only works against some frequencies. [M] According to Sir John Pendry, a professor at Imperial College London – "as with new technologies there is sometimes a little hype when reporting preliminary results." [M] Lending his thoughts and downplaying Chinese research in that the mesh or materials they are using are not exactly new, but new enough that their promise often exceeds their actual performance and usefulness. [M]

Figure 8 Daily Beast Image Depicting Stealth Radar [M]

According to Reuters, Chinese defense spending increased a total of 7.5 percent for 2019 from 2018 averaging a total of 1.19 trillion yuan or $177.49 billion. [H] China is attempting to develop new military capabilities, including stealth fighters, aircraft carriers and anti-satellite missiles. According to Chinese Premier Li Keqiang telling parliament, "[w]e will further implement the military-civilian integration strategy, and speed up efforts to make innovations in defense-related science and technology." [H] In addition to spending, research instances such as a Duke PhD students case, "who allegedly appropriated sensitive research funded by the US Military on metamaterials, and then returned to China to fund a highly successful research institute, the Kuang-Chi Group, which supports the Chinese military in advanced technology developments." [H] Finally, another

Figure 9 Reuters Report on China Investment [H]

Click Picture to be connected to video (must have internet connection)

example is, the China Electronic Technology Group Corporation (CETC) which is a backbone State-owned enterprise in China, [M] which formed a partnership with the "University of Technology Sydney in Australia to establish the Australia-China Research Innovation Centre in Information and Electronics Technologies." [H] The State-owned company CETC will provide as of 2017, $20 million over five years. The joint venture will engage in "research programs focused on big data technologies [such as] mobile sensing and communications, electromagnetic metamaterials and devices…" [H]

Various Chinese academies and institutions are working towards enhancing stealth technology. Ground and space based remote sensing for air pollution research at a Chinese university provided the radiance differences between wavelength and wavenumber spaces in convolving hyperspectral infrared sounder spectrum to broadband for inter-comparison. [H] The study was supposedly conducted for weather elated purposes, but could possibly be enhanced for military application. According to DefenseOne, "tanks must become stealthy to survive on the future battlefield. . . [reducing] their infrared, acoustic, and radar signatures. . . able to blend in with their heat surroundings better, reducing or masking [their] temperature." [H]

China is further emphasizing its studies on modernizing its military "equipment and weapon systems' capabilities, endurance and lifespans." [M] On or about April 2019, Business Insider India, quoting the Chinese Communist controlled tabloid, Global News reported, "China was developing stealth and camouflage materials that could shelter an aircraft or vessel from broad electromagnetic spectrum. . . a new material, which was said to be lighter, more robust and flexible than materials used by the military, could be also used on vehicles and clothing." [L]

There is no data on how much money China is putting into camouflage development, but to state an overall increase of 7.5 percent in 2019. Given that Xi Jinping's vision of a rejuvenated China includes military might, as well as investments from central and local government into/from the private sector should be large.

Analytic Confidence:

The analytic confidence for this estimate is *moderate*. Open sources were used and were generally reliable. Source content varied; however, sources tended to corroborate one another. There was adequate time allotted to the task and the analyst used a structured research method.

Authored by: Ashraf Abdelhak

China's Modernization of Autonomous Capabilities Will Make Frequency, Radio Wave, Visual, and Wavelength Signatures Harder to Detect by 2030

Executive Summary:

Due to their the most aggressive and advanced autonomous military program to date, by 2030, China will reduce their autonomous system signatures of frequency, radio wave (RW), visual, and wavelength making it difficult for adversaries to detect those systems. Despite China being a long-time production-oriented country, it is slowly increasing its domestic development abilities of autonomous systems through innovation incentives and technology theft.

Discussion:

Artificial intelligence is a key underpinning to autonomous capabilities, see China AI report. China is likely to achieve its goal of becoming the world leader in artificial intelligence with the capability of autonomous systems by 2030, according to a new analysis which suggest Chinese researchers are already on the track to reach the goal

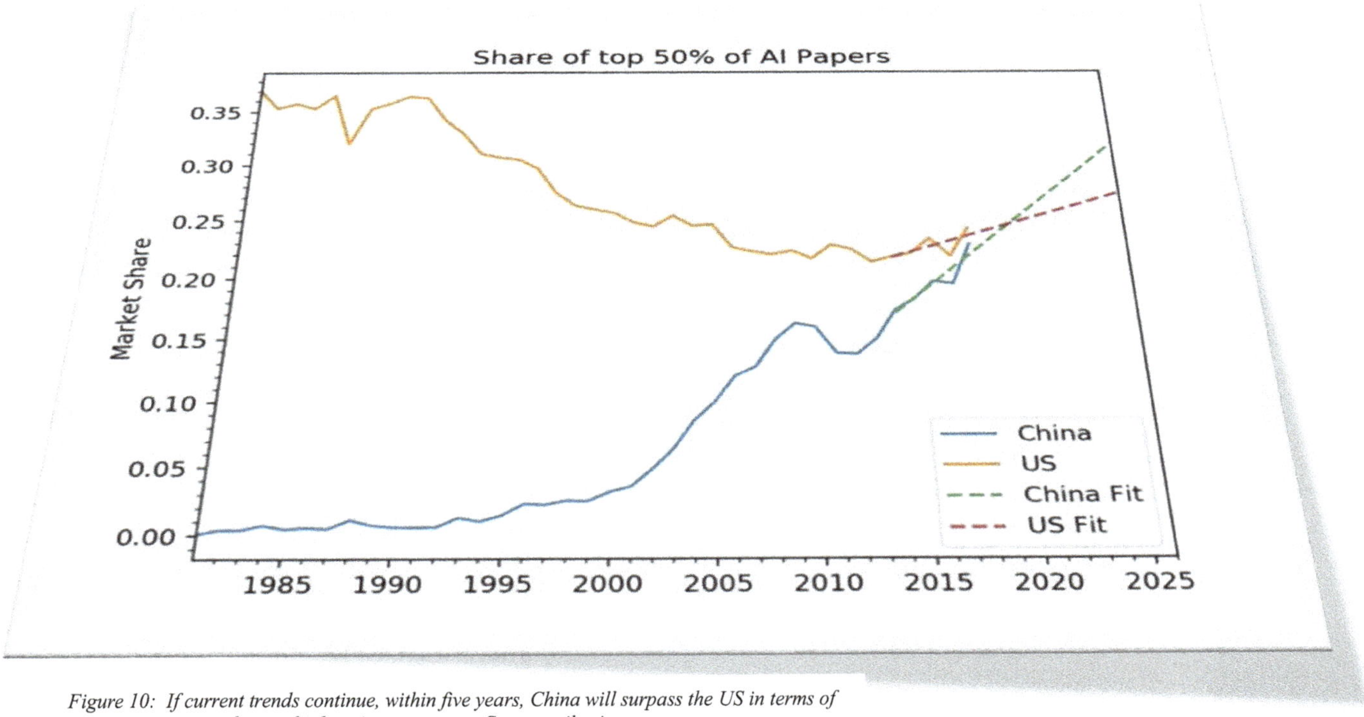

Figure 10: If current trends continue, within five years, China will surpass the US in terms of the top, highest-impact papers. Source: aibusiness.com

within the next five years.[M] The research, conducted by Allen Institute for Artificial Intelligence, a Seattle-based company that focuses on AI specifically, shows that China is set to overtake the US in the most-cited 50% of papers this year, the most-cited 10% of papers next year, and the 1% of most-cited papers by 2025, see figure 10. Striving to achieve a competitive edge, China likely will continue offer incentives such as those in the Artificial Intelligence Development Plan (AIDP) and China's AI Innovation Action Plan. [M]

Figure 11: A Blowfish helicopter drone drops bombs while hovering.
Source: Global Times

In pursuit of advancing military capabilities, China seeks to create "intelligentized" military weapons. In 2018, Major General Ding Xiangrong, Deputy Director of the General Office of China's Central Military Commission, expressed China's goal to take advantage of the "ongoing military revolution… centered on information technology and intelligent technology".[M] By 2030, China will likely connect three of the five domains -- air, land, sea-- with autonomous capabilities.

China's race for autonomous military power has China's large industrial engineering companies like CATIC (China National Aero-Technology Import & Export Corporation) focusing on autonomous technologies. CATIC recently created the Ziyan Blowfish (Figure11), an unmanned helicopter equipped with bombs that "autonomously performs more complex combat missions, including fixed-point timing detection, fixed-range reconnaissance, and targeted precision strikes." [M]

China has offset their UAV detectability by reducing frequency, radio wave (RW), visual, and wavelength signatures by keeping UAV profiles small and AI self-contained, see China UAV insert. The Blowfish UAV smaller profile at 1.87 meters wide, 0.62 meters tall Blowfish is still capable of carrying 12-kilogram payload, such as radar systems, jamming devices, guns or bombs under its spine, and attain speeds of 130 kilometers an hour.[M] See China EW article on radio waves and jamming outputs. China's stealth technology complements reducing a system's signature profile, see China's stealth insert. The US fully expects that the Chinese government will export armed drones that are fully autonomous. Defense Secretary Mark Esper said in a speech aimed at drawing distinctions between the United States and China's competing approaches to artificial intelligence. [M]

Figure 12: The state-owned China prototype of the landing craft The Verge

Crossing both land and sea domains, state-owned China Shipbuilding Industry Corporation (CSIC) is building a prototype autonomous amphibious landing craft, figure 12, that can plot out its own route, swim to shore, avoid obstacles, and also be remotely controlled by an operator. [M] Of interest is the reported (future) ability to integrate with other autonomous systems, like drones or boats. The craft can provide China's Marines with firepower, conduct reconnaissance, or be staged for future use. The difficulty the US faces in tracking signatures of China's autonomous landing craft is its capability to transition from one domain to another sea to land and land back to sea (Figure 13). [M]

Figure 13: China demonstrate the transition of domain from sea to land in a prototype of an autonomous landing craft Source: The Verge

In addition to China aggressiveness in seeking to be the leader in the autonomous war against the US, EW and AI are becoming a more common feature in China joint warfare training exercises, see China EW and AI reports. Without the US mitigation efforts, China's holistic autonomous approach will increase the difficulty for US to detect or identify China's autonomous future weapons.

Analytic Confidence:

Analytic confidence on this analysis is moderate. Open sources were used and the reliability of the sources available on this topic were above average with the several high-quality sources available for the estimate. Source content varied; however, sources tend to corroborate on another. There was adequate time allotted to the task and the analyst used a structed research method.

Authored by: Jerry Brown

China's Mobile, Ground-based, Anti-satellite, Laser Weapon Capability Likely to Make Thermal, IR, Laser, Physical, and Non-Technical Signatures Harder to Detect Within 15 Years

Executive Summary:

Despite directed energy weapons being a lower, national-level priority than other military projects, China continues its technological advancements in ground based, anti-satellite technology efforts. Due to China's previously established and demonstrated laser technologies, it is likely that China will mature existing laser technology to create a mobile, ground-based laser capable of damaging the optics or sensitive components of low Earth orbit (LEO) satellites within the next 15 years. The increased capability likely to make thermal, IR, laser, physical, and non-technical signatures such as unit footprint harder to detect.

Discussion:

Figure 14: Chinese anti-drone, ground-based laser and power source. Source: scmp.com

While destructive laser technology is still largely experimental, technology improvement trends continue to advance the potential damaging effects on satellites. In 2006, China used a ground-based laser to "dazzle" an orbiting U.S. satellite in what was viewed as a capability demonstration. [H] China continues to advance laser capabilities and consolidated research and development operations. China established the Strategic Support Force (SSF) in 2015 possibly for research, development, testing, and fielding of certain "new concept" weapons, such as directed energy and kinetic energy weapons. [H] China also constructed a directed energy weapons facility in China's northwestern province of Xinjiang. [H,M] There have been several, recent Defense agency reports that discuss China's continuing pursuit of anti-satellite weapons, both ground- and space-based. [H] The Defense Intelligence Agency suggests China may field higher power systems that extend the threat to the structures of non-optical satellites. [H]

The physical signature of ground-based lasers capable of damaging LEO satellites require a large power source--much larger than the low altitude, anti-drone laser depicted in figure 14. The People's Liberation Army (PLA) is evaluating the LAG II, seen in figure 15, for anti-drone missions. [M] This laser is powerful enough so that, when linked to a fire control radar, it is likely to also shoot down enemy artillery shells, rockets and missiles. [H] Mobilization of a more powerful, yet similar system suggests further advance technologies for even more powerful lasers are likely. While no sources directly indicate a timeline, the trend of laser and power improvements suggest a capability of damaging LEO satellites in the next 15 years. The platform's mobility and power improvements will likely create difficulty of detection of thermal, IR, laser, physical, and non-technical signatures such as unit footprint.

Figure 15: LAG II Mobile, Ground-based laser potentially capable of shooting down drones, artillery shells, and missiles. Source: globalsecurity.org

Damage to US satellites is largely military significant. Lower powered lasers can temporarily blind satellite optical sensors or detectors on early warning satellites used to detect missile launches, and the electro-optical transducers on electro-optical reconnaissance satellites. [M] An intense laser strike of 300 watts per square centimeter can melt the surface of satellite optical glass and cause optics to fail [M] which would degrade surveillance from satellites. With many US systems reliant on the Global Positioning System (GPS), damage to the critical GPS information would likely negatively impact military operations.

Analytic Confidence:

The analytic confidence for this estimate is *moderate*. Open sources were used and were generally reliable. Source content varied; however, sources tended to corroborate one another. There was adequate time allotted to the task and the analyst used a structured research method.

Authored by: Russell Hoff

China's Frequency, Radio, Physical, and Visual Signatures of Unmanned Aerial Vehicles (UAVs) Will Likely Become Difficult to Detect Within the Next 15 Years

Executive Summary:

China's unmanned aerial vehicles (UAVs) technology improvement is increasing including the incorporation of advanced Command and Control, Computers, Information/Intelligence, Surveillance, Targeting Acquisition and Reconnaissance (C4ISTAR) capabilities which will highly likely surpass U.S. production within the next 15 years. With vast experience in producing UAVs and increasingly better technologies, China's frequency, radio, physical, and visual signatures of UAVs will likely become difficult to detect within the next 15 years. Due to a profitable market, strategic advantages, and the ability to integrate illegally acquired technology, China's UAV capabilities continues to increase with a high production rate capacity. Despite current reliance on foreign technologies, China has opened UAVs facilities to reverse engineer and improve technologies to build their own technology capacity, which likely translates to improved UAV capabilities.

Discussion:

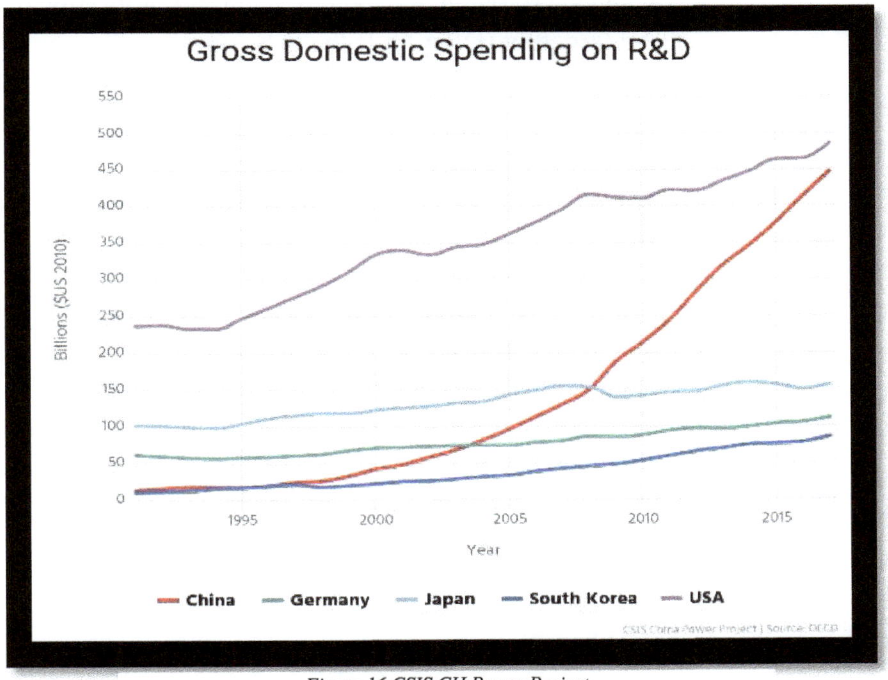

Figure 16 CSIS CH Power Project

With mass production of advanced and reliable UAVs, China is making an unprecedented statement in the global market and continues to demonstrate its ability as a challenging near-peer competitor. Figure 16 graphically depicts the Organization for Economic Co-Operation and Development (OECD) report that "according to some estimates, China will overtake the US as the top R&D spender by 2020." [H] China is working diligently to improve UAV's technological quality and "it is well known that the PLA is seeking to close the technological and capability gap with other leading military powers such as the United States." [M]

With vast experience in producing UAVs and increasingly better technologies, it is likely China's UAVs will have reduced frequency, radio, physical, and visual signatures. China has begun the production of low signature, stealth UAVs such as the CH-7 and Sharp Sword. See China Stealth article for more information on China's stealth capabilities. The CH-7 "under development by the China Academy of Aerospace Aerodynamics (CAAA), the flight technology development arm of China Aerospace Science Corporation (CASC), the prototype CH-7 on display is at the detailed design stage and adopts a truncated diamond-shaped 'flying wing' platform that is equipped with sweptback outer wings with control surfaces let into the trailing edges of the air vehicle's fuselage, and trailing edges and upper surfaces of its wing sections."[13] Future China's UAVs radio signatures likely will be difficult to detect without upgrades of the US Army's current active electronically scanned array (AESA) which is designed to detect larger equipment than small UAVs. The smaller the object, the more difficult it is for radar systems, AESA included, to detect and identify the UAVs' signatures. [H] Additionally, autonomous UAVs imbedded with AI will likely reduce detectable radio signals emissions since commands would likely be self-contained within the platform, see China autonomous statement, see China AI report.

The U.S. is the leading authority in UAV technology and must continue to conduct extensive research and development in UAVs signatures detection, recognition, analysis, and targeting (DRAT). If the U.S. significantly improves its UAV signature detection capability, it is likely that China's future UAVs signatures will revert to a moderate detection challenge. U.S. ground tactical military units are employing a variety of mounted and dismounted UAV's signatures detection devices already equipped and prepared to conduct DRAT operations to counter China's UAVs. China's latest UAV technology and production is advancing, but not yet comparable to that of the U.S. and other near-peer technological advancements. [M] However, new emerging small size UAVs signatures including frequency, physical, radio, and visual are likely to improve in China's future UAVs development. In addition to the frequency, radio, physical, and visual signals, the electro-optical (passive optical) and thermal signatures (passive thermal) are other top components of a long list of advanced UAVs signatures and "the advantages of the former [passive visible imaging] include the possibility to distinguish between a UAV, birds, or any other object in the sky fairly inexpensively due to commercial availability. The advantages of the latter [passive thermal imaging] are the reduction of background signal noise, enhanced nighttime detection and less susceptibility to weather degradations. Conversely, disadvantages include the need for a clear unobstructed view of the object and high susceptibility to background signal noise

[13] "Airshow China 2018: CASC reveals stealth UCAV development" UK, Jane's International Defence Review 06 Nov 18 (accessed through Jane's Group UK Limited): "China Aerospace Science and Technology Corporation (CASC) used Airshow China 2018, held in Zhuhai from 6–11 November, to announce it is developing a turbofan-powered high-altitude unmanned combat air vehicle (UCAV) called the Cai Hong 7 (Rainbow 7, or CH-7). Under development by the China Academy of Aerospace Aerodynamics (CAAA), the flight technology development arm of CASC, the prototype CH-7 on display is at the detailed design stage and adopts a truncated diamond-shaped 'flying wing' planform that is equipped with sweptback outer wings with control surfaces let into the trailing edges of the air vehicle's fuselage, and trailing edges and upper surfaces of its wing sections."

of the former [passive visible imaging] and the need to overcome the act that most UAVs have low thermal signatures of the latter [passive thermal imaging]." [H]

It is likely that future detection of China's UAVs acoustic signatures will remain moderately detectable as Busset, et al, 2015 described "despite the advantages of being a passive sensor that is relatively inexpensive, acoustic sensors would make inefficient standalone detection systems." [H] Usually, acoustic signatures are limited to specific UAVs signatures relying on the recognition of specifics UAVs characteristics.

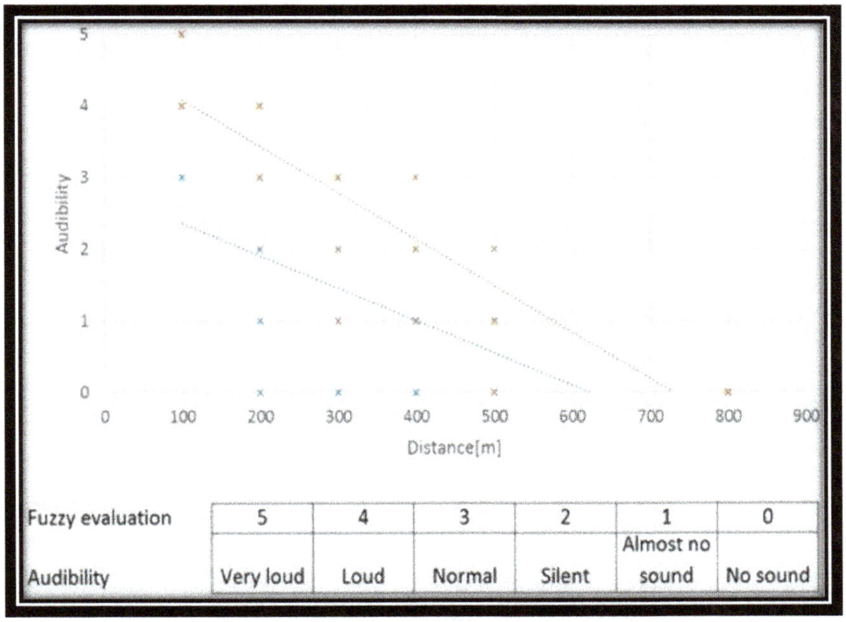

Figure 17 Visibility and Distance Matrix (Blue: Sky; Orange: Urban; Grey: Dry Grass

The least effective method of UAV signature detection is human sensing. China's future UAVs will likely increase in speed and lethality making the signature recognition though electronic means more critical for defensive and offensive operations. Experiments on the efficiency of the human hearing as a source of UAVs signature detection demonstrated that time, precision, and discrimination of UAVs is critical and required advanced technical instrumentation. Figure 17 from the Department of Air Defence, Faculty of Military Technology, University of Defence, Czech Republic, depicts the relationship of human hearing and distance and concludes that at 500 meters and altitude of 50 meters human hearing is ineffective for the average medium size commercial UAV. Blue represents a 7-10m altitude, and orange represents a 50-100m altitude, and notes that a human's effective hearing detecting a medium size UAV is at an average of 300m. [H]

China UAVs Future Capabilities:

China's UAV weapon payloads are rapidly increasing in lethality and distance of travel. A recent assessment on the development of a unmanned carrier aerial vehicle (UCAV) demonstrated China's aggressive efforts to technologically surpass all competitors. [M] As China's UAV technology and production improves, the UAV improvements result reduced frequency and radio wave signatures. In the book, China's Strategic Modernization: Implications for the United States, Mark Stokes confirms China's UAVs advanced technology making improvements on global positioning systems which emits radio waves signatures while simultaneously reducing the low radar cross-section signatures. [H] However, UAVs' signature reductions is not the primary aim in China's modernization. China is not making a major effort to produce UAVs with impressive signature detection avoidance. The main reason of why China's UAV signatures will be categorized as difficult to detect rather than very difficult is due to "the 'People's Liberation Army's (PLA) modernization strategy is still driven by the need to close the gap it sees with other peer forces (chiefly the United States) in conventional power. As a result, the highest emphasis is being place on legacy platform development for air, land and sea: better aircraft (manned), tanks, and ships." [14]

The U.S. recently developed its first UCAV platform but is behind China's efforts due to other priorities. [M] When comparing China and US UCAV's technology, Elsa Kania, a fellow at the Center for New American Security, said that "the drone might even allow China to partially catch up to the U.S. in carrier-based combat." [H] Furthermore, a U.S. Department of Defense (DOD) annual report to Congress emphasized China's rapid UAV production and marketing where "the Middle East and North Africa region was China's second largest regional arms market, probably because of the demand for armed UAVs – a niche are where China is one of the few supplier." [H]

While China aggressively provides armed UAVs to the Middle East and North Africa regions, DOD confirmed that the U.S. is losing technological ground as "the Peoples Liberation Army Air Force (PLAAF) continues to modernize and is closing the gap with the U.S. Air Force across a broad spectrum of capabilities, gradually eroding the United States' longstanding significant technical advantage." [H] _The aggressive construction of UAV plants demonstrates China's initial efforts to surpass the U.S. in UAV technology and production "as reflected by the inclusion of a number of unmanned systems technology R&D projects under China's National High Technology Research and Development Program, also known as the 863 Program." [H]

While China dedicates approximately 60 billion of fiscal resources in accordance with the China Power Project (CPP), [H] "the development of unmanned systems appears to receive considerable national-level funding and support in China." [H] Counter to the U.S. current lead in UAV technology, the Global Security Organization (GSO) noted the potential for China closing the UAV technological gap with the U.S. because of large funding, and China's understanding of the strategic advantage of UAVs mass production. GSO specifically focused on the budget as the catalyst for research and development explaining that "China's rapidly expanding defense budget supported impressive

---

[14] "Unmanned dragons: China's UAV aims and achievements" UK, International Defense Review, 23 Jan 12 (accessed through Jane's Group UK Limited): "There are several reasons why China is still some way from the cutting edge of UAV technology and employment. For one thing, the People's Liberation Army's (PLA) modernization strategy is still driven by the need to close the gap it sees with other peer forces (chiefly the United States) in conventional military power. As a result, the highest emphasis is being placed on legacy platform development for air, land and sea: better aircraft, tanks and ships. Within PLA strategy cells and think thanks, much time is devoted to future asymmetric combat concepts that would include UAVs and other disruptive systems, but the PLA clearly wants to push through a conventional force modernisation before moving on to this next phase."

advances in drone technology, prompting some to worry that the United States' global dominance in the market could soon be challenged." H

Another reason for the moderate improvement on reducing the signatures is that China's UAVs mass production is connected to the suspected illegal acquisition of UAV technology and limited reversed engineering. As evidenced during the U.S. House of Representatives Committee on Armed Service, the Honorable Michael D. Griffin, Undersecretary of Defense for Research and Engineering, confirmed that "China has made it a national goal to acquire foreign technologies to not only advance its economy, but also to use these technologies to advance its military capabilities, and it is doing so through both licit and illicit means." H China's UAV reverse engineering began with the purchase of U.S. products from allies such as Israel and Japan providing China with the ability to acquire advance US UAV technology. H China's UAV development infrastructure is capable of reverse engineering adding concern to the U.S. RQ-170 drone lost in Iran, but "even if the Chinese aerospace industry can't use reverse engineering to produce an indigenous equivalent of the RQ-170, Chinese engineers could probably learn enough from the RQ-170 to develop improved countermeasures and defenses against it and similar systems." H

Equally important to China's ability to integrate illegally acquired technology is the manipulation of exportation laws showing that "indeed, China could be well positioned to become the top global source for many countries seeking such systems; some reports state that China will produce over half the world's UAVs by value over the next decade." H A Global Security report affirmed that "China is said by Chinese sources to be the largest exporter of military drones. The best-known Chinese military drones are the Wing Loong family, made by Aviation Industry Corp of China, and China Aerospace Science and Technology Corp's CH series." H

Finally, China is exploring the strategic advantages of employing UAVs Swarms to achieve national level objectives. The employment of UAVs swarms could likely increase the signature of the UAVs, but give the employers the advantage of mass confusion, deception, and psychological dominance. Currently, China has already experimented and "set a world record in December 2017 at the Global Fortune Forum in Guangzhou when it succeeded in mobilizing the largest swarm of drones in history. Over 1,000 miniature drones performed a variety of tasks to showcase the collective orchestration of the high-tech instruments." H The same year in 2017, China's National University Technology (NUDT) conducted several experiments employing UVAs swarm to test strategic capabilities. An interesting finding during China's UAVs swarm employment was the ability for each UAV to operate with its own mission and autonomy from the rest of the UAVs (See China autonomy insert). Even more strategic, is the combat power the UAVs swarm achieves when "future swarms of small drones might also be able to carry electronic warfare jammers, emitters that mimic the signals of larger aircraft, equipment capable of conducting cyber-attacks, or other systems to confuse or overwhelm an opponent's defenses ahead of or during a more complex operation." [H]

Analytic Confidence:

Analytic confidence on this analysis is moderate. The task to find information on China's intention and progress on technology was complex, and this estimate is sensitive to rapidly emergent information. The reliability of the sources was average. All sources presented related thesis supporting this analysis, and China's latest UAVs demonstrations and sales reinforce the validity of sources.

Authored by: Rafael Duran

China Likely to Weaponize AI by 2030; Enabled AI Systems Likely Diminishes Electronic, Digital, and Metanalytic Signatures

Executive Summary:

Due to China's focus on Artificial Intelligence (AI) and expected investments from state, local, and private investors, China is likely to weaponize AI to aid military with intelligence analysis, decision-making, vehicle autonomy, weaponry, and information operations by 2030. Military systems augmented with AI likely to diminish electronic, digital, and metanalytic signatures making those systems more difficult to detect. Despite China's challenges with retaining its AI talent base, China's close ties between its government and private sector firms enable China's defense and intelligence industries to focus AI research and implementation using China's vast collection of data.

Discussion:

Figure 18. Chinese domestic surveillance using facial recognition software. Source: MercatorNet.com

China's leadership has a national focus on AI development. On July 20, 2017, China issued its Next Generation Artificial Intelligence Development Plan (AIDP) that details a top-level design emphasizing the nation's AI goals to become the world leader in AI by 2030.[M] China's president, Xi Jinping, reinforced China's AI priority during his remarks at a 2018 Politburo study session that China must "ensure that our country marches in the front ranks where it comes to theoretical research in this important area of AI, and occupies the high ground in critical and AI core technologies." [M]

Since 2015, China has seen a boon in national AI talent base in both universities and defense institutions; however, about 85% of the talent left China for America to work at tech giants such as Google and IBM.[M,M] There is no official data on how much money China is putting into AI development; however, at least two Chinese regional governments have each committed to investing 100 billion yuan ($14.7 billion) focusing on AI industry, [M] including the major port city of Tianjin. [M] The Chinese Institute of

Electronics [M] and the iResearch Consulting Group [M] each released reports in 2018 indicating China's AI industry investments to be between $20-22 billion by 2022. Despite the talent drain or high talent turnover, China's close ties between its government and private sector firms enable China's defense and intelligence industries to focus AI research and implementation using China's vast collection of data. At the time of this report, China has 19 institutions of higher learning including defense universities and 8 major civilian firms working on aspects of AI.[H,H] In an effort to gain, retain AI talent or reverse the brain drain, China likely will continue offer incentives such as those in the AIDP and China's AI Innovation Action Plan.[M] "Research in AI technologies in Chinese universities is now catching up with their top-class peers worldwide. For some AI techniques, the gap between Chinese universities and research institutions and other AI giants is becoming smaller and smaller; and for some functional applications, we have made great progress and may be a few steps ahead of other key players," said Haifeng Wang, senior vice president at Chinese internet company Baidu. [M]

China current use of AI is widespread. China routinely uses facial recognition for shopping, paying, and accessing some public services. [M] Recent AI advances have made it possible to identify individuals not only in up-close still photos, but also in video—a far more complex scientific task. [M] China also uses AI-driven facial recognition to catch jaywalkers, and "social credit" – an AI-driven credit score that factors in desired social behaviors – are already in use [M] (See figure 18).

Figure 19: The video shows state-owned a China Electronics Technology Group
Click Picture to be connected to video (must have internet connection)

Just as oil fueled the industrial age, data is fueling advances in the AI age. As *The Economist* stated, China is "the Saudi Arabia of data." [H] The chief advantage China has over the U.S. is the access to troves of data from online commerce and social networks afforded by its scant privacy protections. [H] China demonstrated its ability to exploit collected data either in a seemingly innocuous form of face-swapping [H] or more sinister application to identify and intern Muslim ethnic minorities in mass detention camps. [M]

Commercial AI developments, using the data to train AI machines, can then be leveraged into military technology. Gregory Allen with the Center for a New American Security stated that, "[China's] hope is that they will have an advantage in adopting AI technology, and they will use that to build a military that is stronger than that of the United States." [M] China's experience using data could likely be used for machine learning in military applications such information processing and information analysis to which ultimately lead to decision-making abilities.

Figure 20: AI deepfake disinformation video. Source: YouTube BuzzFeed Video

Click Picture to be connected to video (must have internet connection)

China's military "is funding the development of new AI-driven capabilities" in battlefield decision-making and autonomous weaponry, says Elsa Kania, a fellow at the Center for a New American Security in Washington, D.C.[M] Coupling AI with unmanned systems is likely to enable advancement in swarming capabilities. [M] See figure 19 for emerging swarming capability demonstration. See China UAV article for additional information on swarming. Coupling AI with cyber will likely advance cyber-attacks through the Internet of Things, see China Cyber insert. Deep fake technology exists today, see figure 20. China likely can use AI and machine learning to create deep fakes of deceiving content as a disinformation tool during information operations. [M]

China's focus on AI, infusion of research investments, current AI implementation, and vast collections of data for AI improvements indicate that China is likely to continue technological development with a focus on modernizing its military's abilities. It is likely that processors with decision-making algorithms, enabled by machine learning, will be pre-programmed, self-contained on "smart" platforms, see China autonomous insert, it is likely that legacy detection methods for electrical, digital, or metanalytic signatures will be more difficult.

Analytic Confidence:

The analytic confidence for this estimate is *moderate*. Open sources were used and were generally reliable. Source content varied; however, sources tended to corroborate one another. There was adequate time allotted to the task and the analyst used a structured research method.

Authored by: Russell Hoff

China Highly Likely to Reduce Offensive Cyber Signatures by 2030; Detection, Attribution Highly Likely More Difficult

Executive Summary:

Due to China's military modernization efforts towards meeting its stated, national goals, China is highly likely to improve offensive cyber capabilities by 2030, which includes reducing cyber signatures – electronic, digital, metanalytic -- with AI. Despite theorized shortcomings, such as improperly taught AI, China continues improving cyber espionage and cyber-attacks.

Discussion:

By 2030, China is highly likely to employ the weaponized use of AI (see China AI article), which includes machine learning, to improve offensive cyber capabilities to increase attack successes while reducing common cyber signatures. Unclassified, open source reports indicate that adversaries have not yet used AI for offensive cyber, or possibly not used AI to its fullest potential. [M,M,M] Theoretically, an improperly taught AI system or an AI system that learns incorrectly can result in unwanted outcomes. [M] Additionally, AI powered cyber-attacks might reveal themselves should the number of attacks increase significantly on a system as the unsupervised AI learns, adapts, and re-engages the system's defenses. [M] Despite some theorized shortcomings, China is investing heavily in, learning from, and improving AI applications (see China AI insert) which will lead to inevitable use in cyber-attacks.

While not an all-inclusive list, signatures of cyber-attacks include metadata, [H] "signatures" within code; [H] digital residue, [M,H] pirated software flaws, and digital manipulation (such as anomaly and misuse).[H] Metadata includes document-clustering such as commonly used malware, delivery mechanisms, and infrastructure.[15] "Signatures" within codes include the malware author's unique tag [H] or digital residue includes patterns or artifacts left by the software tool used to create the malware. [M,H] AI powered malware, if properly taught, can likely avoid tell-tale signatures.

[15] "FireEye aims ATOMICITY model at cyber intelligence analysis ", Jane's International Defence Review 04-Jun-2019 (assessed through USAWC Subscription Database): "[Cyber security experts at FireEye] said its ATOMICITY model uses a document-clustering approach, grouping together vast quantities of data on how attackers operate, including commonly used malware, delivery mechanisms, infrastructure, and more. With machine learning techniques, FireEye can then detect patterns across this data so it can establish whether an attack is being pursued by a known group, said Benjamin Read, senior manager for cyber espionage analysis." Source Reliability: High.

China continues to steadily improve its cyber capabilities as expressed in published, national strategies (Figure 21). China's 2013 Science of Strategy contends that the PLA should integrate space, cyber, and electronic warfare operations to "paralyze enemy operational systems."[H] "China's Military Strategy" published in May 2015 identified cyberspace as one of four "critical security domains" and stated the government's intention to speed up the development of a cyber force to tackle "grave security threats" to China's cyber infrastructure. [H] The same document revealed the country will "expedite the development of a cyber force" and enhance "support for the country's endeavors in cyberspace." [H] Likely pursuant to the 2013 strategy, China created the Strategic Support Force (SSF) in December 2015 to integrate the PLA's space, cyber, and EW domains, to enabling multi-domain war-fighting capabilities, to fight and win future informatized wars.[H] The Network System Department of the SSF likely is the convergence of PLA's capabilities for cyber, electronic, and psychological warfare into a single force which likely enable it to take advantage of key synergies among operations in these domains.[16]

Figure 21: Representation of Chinese military emphasis on cyber. Source: https://nsarchive.gwu.edu/news/cyber-vault/2019-01-17/cyber-brief-chinas-military-use-cyber

[16] "The Strategic Support Force and the Future of Chinese Information Operations", Army Cyber Institute, The Cyber Defense Review, Vol. 3, No. 1 (SPRING 2018), pp. 105-122 (accessed through USAWC Subscription Database): "Concurrently, the SSF has integrated the PLA's capabilities for cyber, electronic, and psychological warfare into a single force within its Network Systems Department (网络系统部), which could enable it to take advantage of key synergies among operations in these domains. However, beyond the SSF, the PLA also appears to be building up network-electronic operations (网电作战) capabilities within its national Joint Staff Department headquarters and within new regional theater com-mands (战区), reflecting the emergence of a multi-level force structure specializing in information operations. Thus, the SSF reflects the PLA's uniquely integrated approach to force structure and operations in these vital new domains. This realization of this paradigm through the SSF will enhance the PLA's capabilities to fight and win future "informatized" (信息化) wars." Source Reliability: High.

The US-China Economic and Security Review Commission released an October 2018 report indicated Chinese institutions conducting security research on "attacks that take advantage of heterogeneous, dynamic, and decentralized Internet of Things (IoT) networks with multiple connected devices" which include leveraging machine learning techniques that are "particularly useful in an offensive context."[17] The report added that there was a "high degree of collaboration" between academia, the private sector, and government and military organizations in Chinese IoT security research, claiming that China's "coercive apparatus supervises and directs the collection and release of vulnerabilities", and that "this practice of stockpiling vulnerabilities for exploitation extends to IT products sold in US markets".[18] The US is concerned that either Chinese aircraft or UAVs sourced with Chinese components could be used to send intelligence back to China on US operations. It is possible that China could use its UAV companies to collect data because the Chinese government supports these companies with funding and engineers.[19]

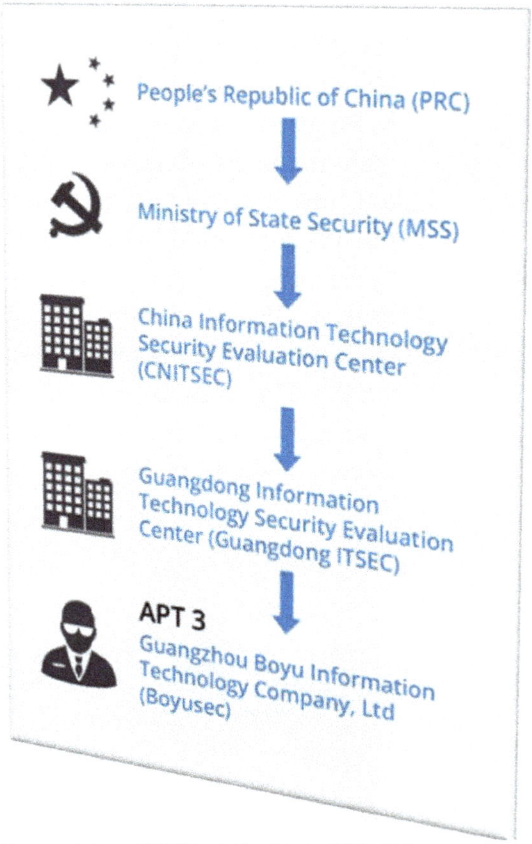

Figure22: Representation of APT3 relationship to MSS. Cyber-security groups also attribute APT-10 & APT-17 to China. Source:https://www.recordedfuture.com/chinese-mss-behind-apt3/

[17] "Expanding IoT poses cyber-security risks and intelligence collection opportunities", Jane's Intelligence Review 30-May-2019 (accessed through USAWC Subscription Database): "An October 2018 report released by the US-China Economic and Security Review Commission provided an insight into the range of institutions conducting security research into the IoT in China. The report noted a dramatic expansion in papers being published by Chinese security researchers between 2009 and 2017, up from nine to 1,229 by one metric, as well as a focus in the papers on "attacks that take advantage of heterogeneous, dynamic, and decentralised IoT networks with multiple connected devices". The authors singled out the work of Chinese IoT security researchers to develop "algorithmic and machine learning techniques for the discovery of IoT vulnerabilities across a wide range of devices", which they described as "particularly useful in an offensive context". Such research would be relevant to developing software systems that could autonomously identify vulnerabilities across the large attack space of the IoT and pivot between different classes of devices. The report added that there was a "high degree of collaboration" between academia, the private sector, and government and military organisations in Chinese IoT security research, claiming that China's "coercive apparatus supervises and directs the collection and release of vulnerabilities", and that "this practice of stockpiling vulnerabilities for exploitation extends to IT products sold in US markets".

[18] Ibid.

[19] "US bans Pentagon use, procurement of Chinese UAVs", Jane's International Defence Review 30-Dec-2019 (accessed through USAWC Subscription Database): "The Department of Defense has purchased several Chinese DJI aircraft over the years, but soldiers are not supposed to be using them. There are also few Chinese C-UAV products, and the US is not buying those. Jane's believes the issue with components is much trickier because there are many systems that could use Chinese parts. There is a concerted effort to develop systems without Chinese equipment or to make the systems more easily modifiable to remove Chinese parts and replace them with parts from non-Chinese suppliers. The US is concerned that either Chinese aircraft or UAVs sourced with Chinese components could be used to send intelligence back to China on US operations. It is possible that China could use its UAV companies to collect

China's 2019 National Defense in the New Era continues to echo a technology focus through the "application of cutting-edge technologies such as artificial intelligence (AI), quantum information, big data, cloud computing and the IoT is gathering pace in the military field." H Acknowledgement of the future strategic importance of cyber warfare, China noted that, "War is evolving in form towards informationized warfare, and intelligent warfare is on the horizon." H Chinese cyber-attacks have been carried out against space organizations as part of the plans for future anti-satellite attacks. M Additionally, EW and cyber warfare are becoming a more common feature in PLA exercises to replicate complex electromagnetic spectrum conditions.[20] Chinese PLA are likely to modernize special forces with the next-generation technology available to conduct the full spectrum of future special operations.[21]

Outside of the PLA organization, in July 2019, US cyber-security groups tied another Active Persistent Threat (APT), APT-17, to China's Ministry of State Security (MSS) (Figure 22) indicating China's APT threat actors likely to continue to evolve and launch new campaigns. M

Chinese cyber personnel likely obtained insights to NSA and CIA cyber tools and techniques that Edward J. Snowden M and, allegedly, Joshua A. Schulte M placed into the public realm. Additionally, incorporating AI highly likely increases Chinese cyber successes by increasing attacks methods and decreasing signature detection.

Analytic Confidence:

The analytic confidence for this estimate is *moderate*. Open sources were used and the reliability of the sources available on this topic were above average with several high-quality sources available for the estimate. Source content varied; however, sources tended to corroborate one another. There was adequate time allotted to the task and the analyst used a structured research method.

Authored by: Russell Hoff

data because the Chinese government supports these companies with funding and engineers. Jane's believes the Pentagon fears China as a near peer competitor more than it does Russia because of the Chinese philosophy to use its entire set of resources of the state, a military/civil fusion, to achieve a goal.

[20] "Jane's Sentinel Security Assessment - China And Northeast Asia: China – Army – Military Exercises", in Jane's (accessed through USAWC Subscription Databases): "[Chinese] Cyber and EW forces are playing a more prominent role in exercises too. The PLA's first major wargames involving cyber forces were held at Zurihe in Inner Mongolia in June 2013, and they included SOF, army aviation, and EW and digital units. EW and cyber warfare are becoming a more common feature in PLA exercises to replicate complex electromagnetic spectrum conditions." Source Reliability: High.

[21] "Jane's Amphibious and Special Forces: Capability Analysis", in Jane's (accessed through USAWC Subscription Databases): "The emergence of the SSF provides an interesting glimpse into the future roadmap of the Chinese SOF. Acknowledgement of the future strategic importance of cyber warfare and EW in particular shows how Chinese SOF are seeking to attain all the next-generation technology available to conduct the full spectrum of future special operations. However, integration into the various Special Operations Brigades and its subsequent connectivity across not only SOF, but also conventional units, will decide upon the degree to which it is used across the contemporary and future operating environments as the PLA continues to seek influence beyond its borders." Source Reliability: High.

RUSSIA

Russia's Estimative Key Findings

Authored By: Samuel Smith

Based on an analysis of Russian modernization plans, defense spending and stated goals that evaluated 19 capabilities against 22 possible signatures, it is highly likely that three signatures—including non-technical signatures such as doctrine, electro-optical signatures, and signatures associated with software changes—will both impact the most capabilities and become harder, on average, to detect in the 2030-2040 time frame (See figure 23 below).

Estimated Signature and Capability Threat Matrix - Russia

Capabilities / Signatures	Electronic Warfare	Deception	Cyber	C4ISR	Hypersonic	Air Defense	Missiles	Tanks	Stealth	Autonomous	Artificial Intelligence	Unmanned Systems	Batteries	Anti-Satellite	Long-Range Strike	Quantum	Chemical	3D Printing	Short-Range Strike	
Electronic (Digital/Software)	4	4	4	4	3	4	4	4	3	3	4	3	3	3	3	3	3	3	3	3.42105263
Visual	4	4	4	4	4	2	3	5	5	4	3	3	3	3	3	3	3	3	2	3.42105263
Non-Technical (Process/Doctrine)	5	5	4	4	4	4	2	3	3	3	3	4	3	3	3	3	3	3	3	3.42105263
Digital	4	4	4	4	3	3	4	3	3	3	4	4	3	3	3	3	3	3	3	3.36842105
Radio waves	4	4	4	4	4	4	4	3	3	3	3	3	3	3	3	3	3	3	3	3.36842105
Speed	4	4	4	4	4	3	3	4	4	4	3	2	3	3	3	3	3	3	3	3.36842105
Magnetic	4	4	4	4	3	3	4	4	3	3	3	3	3	3	3	3	3	3	3	3.31578947
Physical	4	4	4	4	3	4	3	3	3	4	3	3	3	3	3	3	3	3	3	3.31578947
Frequency	4	4	4	4	4	3	4	3	3	3	3	4	3	3	3	3	3	3	2	3.31578947
Thermal	4	4	4	4	4	4	2	4	4	3	3	3	3	3	3	3	3	3	2	3.31578947
Acoustic	4	4	3	3	3	3	4	3	4	3	3	3	3	3	3	3	3	3	3	3.21052632
Hyperspectral	4	4	4	4	3	2	4	3	3	3	3	3	3	3	3	3	3	3	3	3.21052632
Radiation	4	3	3	3	3	4	4	3	3	3	3	3	3	3	3	3	3	3	3	3.15789474
Wavelength	4	4	4	4	3	3	3	3	3	3	3	3	3	3	3	3	3	3	2	3.15789474
x-Ray	4	3	3	3	3	3	3	3	3	3	3	3	3	3	3	3	3	3	3	3.05263158
Laser	3	3	3	3	3	3	3	3	3	3	3	3	3	3	3	3	3	3	3	3
Smell	3	3	3	3	3	3	3	3	3	3	3	3	3	3	3	3	3	3	3	3
Metanalytic	3	3	3	3	3	3	3	3	3	3	3	3	3	3	3	3	3	3	3	3
Ultra Violet	3	3	3	3	3	3	3	3	3	3	3	3	3	3	3	3	3	3	3	3
Seismic	3	3	3	3	4	3	3	2	3	3	3	3	3	3	3	3	3	3	3	3
Infrared	4	2	2	2	3	4	2	4	3	3	3	3	3	3	3	3	3	3	3	2.94736842
Bio/Chemical	3	3	3	3	3	3	3	3	3	3	3	3	3	3	3	3	3	2	3	2.94736842
	3.77272727	3.59090909	3.5	3.5	3.31818182	3.22727273	3.22727273	3.27272727	3.22727273	3.13636364	3.09090909	3.09090909	3	3	3	3	3	2.95454545	2.81818182	

Future Signature Detection Scale (educated guesses)

5 Very Difficult
4 Difficult
3 Moderate to Routine
2 Easy to Detect
1 Extremely Easy to Detect (Signatures Avail via Commercial means)

Figure 23: Russia Estimated Signature and Threat Capability Matrix

In fact, over 21% (89/418) of future Russian threat signatures distributed over 12 of the 19 capabilities will likely become harder to detect between 2030 and 2040. 1% (4/418) will become much more difficult to detect and only 3% (15/418) of the signatures associated with these 19 capabilities are likely to become easier to detect in the same period. The future Russian threat signatures most difficult to detect in 2030-2040 is likely the non-technical (tactics/process/doctrine) signatures from its EW and Deception capabilities/program.

1. Driving these changes are a number of key facts. Specifically:
    a. Based on a report from the Centre for Economics and Foreign Policy Studies and published as part of the NATO at 70 project advances in digital technology have made it possible for Russia and modern militaries to develop highly and adaptable electronic systems with feedback that enable rapid adaptation to the electromagnetic environment.
    b. According to Roger McDermott a senior research fellow in Eurasian Military Studies at King's College in London due to the hybrid warfare, cyber warfare and information warfare lessons learned from 2014 to date in Ukraine and in Syria by Russia military operations, new requirements for Russian equipment development have emerged. Russian President Putin has clearly pointed out, "Russia must put the development of digital technology equipment, artificial intelligence, drones and robot systems on its agenda in terms of Russian military construction".
    c. According to The State Armaments Program for 2018-2027, Russia will invest 19 trillion rubles (about 306 billion US dollars) in the next 10 years in defense procurement and equipment upgrades. Specifically, Russia is planning to develop and procure high-precision weapons for air, sea and land battle – including hypersonic missiles – unmanned air strike complexes, individual equipment for servicemen and advanced reconnaissance, communication and electronic warfare systems in the next 5-10 years.

Type of weapons system	2013	2015	2017	2020
Submarines	47	51	59	71
Surface ships	41	44	54	71
Aircraft	23	37	55	71
Helicopters	39	63	76	85
Ground missile systems	27	64	100	100
Artillery	51	53	59	79
Armoured vehicles	20	37	56	82
Multi-role vehicles	40	48	56	72

Source: Ministry of Defence of the Russian Federation (undated), 'Plan deyatel'nosti na 2013–2020 gg' [Action Plan for 2013–2020], Moscow: Ministry of Defence, http://mil.ru/mod_activity_plan/constr/vvst/plan.htm (accessed 21 Feb. 2018).

Figure 24: Reported share of modern military equipment in total inventories for selected categories of weapons system, per cent (2013-17 = actual; 2020 = target) Source: Chatham House The Royal Institute of International Affairs.

d. According to Russian military experts from published Rand and Hudson Institute reports the introduction of unmanned EW systems as a mean to advance an EW combat operation on to enemy territory is a stated Russia objective and will likely advance this capability by 2030.
e. Based on a report by the Republic of Estonia Ministry of Defense more than 120 Russian companies are involved in the development and production of EW systems.
f. According to Russian military experts from published Rand and Hudson Institute reports based on Russian data Information and communication technology (ICT) promotion is one of the Russian policies' priority areas. Offensive cyber is playing a greater role in conventional Russian military operations and may potentially play a role in the future in Russia's strategic deterrence framework. Although the Russian military has been slow to embrace cyber for both structural and doctrinal reasons, Russia has signaled that it intends to bolster the offensive as well as the defensive cyber capabilities of its armed forces. Hardware continues to dominate the Russian IT market, making up more than 50 percent of the sector.

g. According to Talal Husseini a technology analyst Russia is increasing its use of Radar Absorbent Materials (RAM) and other adaptive camouflage, or active camouflage technology to reduce and conceal threat signatures.

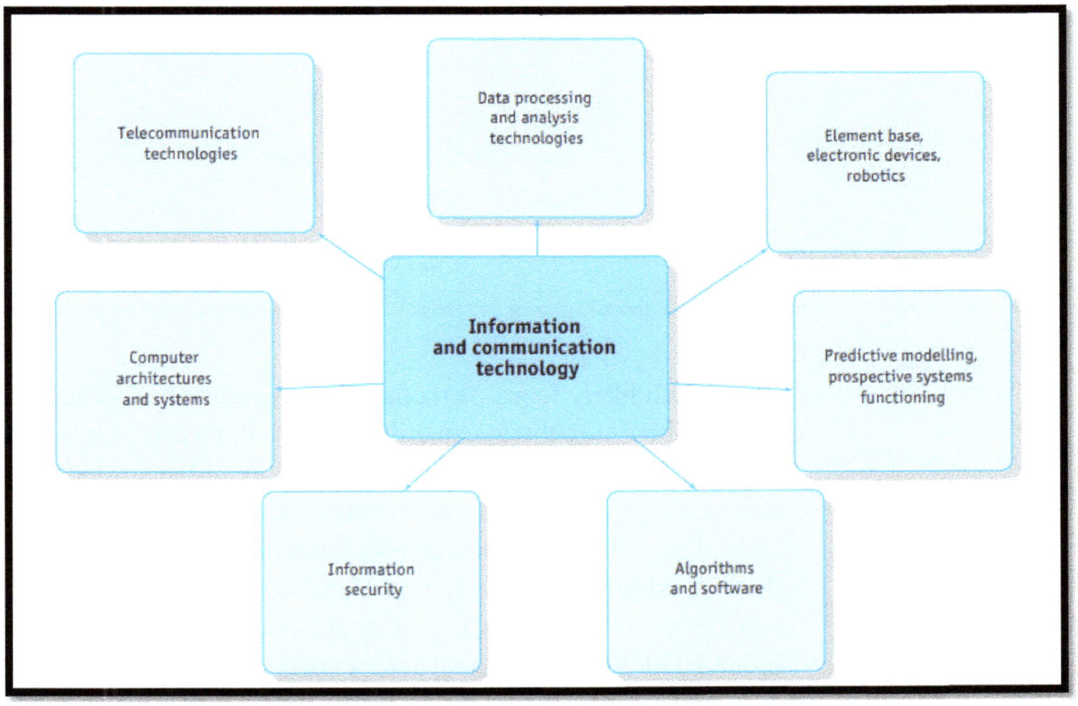

Figure 25: Russian Thematic Fields of the Information & Communication Technology Priority Area. Source: Ministry of Education, Science of the Russian Federation.

h. According to Russia's State Armaments Program for 2018-2027 Russia has developed nuclear hypersonic capability on intercontinental-range missiles presenting challenges for electro-optical, speed and non-technical signature detection.

Estimated Signature and Capability Threat Matrix - Russia

	Air Defense	C4ISR	Cyber	Deception	**Electronic Warfare**	Hypersonic	Missiles	Stealth	Tanks	AVG
Acoustic	3	3	3	4	4	3	4	4	3	3.44
Bio/Chemical	3	3	3	3	3	3	3	3	3	3.00
Digital	3	4	4	4	4	3	4	3	3	3.56
Electronic (Digital/Software)	4	4	4	4	4	3	4	3	4	**3.78**
Frequency	3	4	4	4	4	4	4	3	3	3.67
Hyperspectral	2	4	4	4	4	3	4	3	3	3.44
Infrared	4	2	2	2	4	3	2	3	4	2.89
Laser	3	3	3	3	3	3	3	3	3	3.00
Magnetic	3	4	4	4	4	3	4	3	4	3.67
Non-Technical (Process/Doctrine)	4	4	4	5	5	4	2	3	3	**3.78**
Physical	4	4	4	4	4	3	3	3	3	3.56
Radiation	4	3	3	3	4	3	4	3	3	3.33
Radio waves	4	4	4	4	4	4	4	3	3	**3.78**
Seismic	3	3	3	3	3	4	3	3	2	3.00
Smell	3	3	3	3	3	3	3	3	3	3.00
Metanalytic	3	3	3	3	3	3	3	3	3	3.00
Speed	3	4	4	4	4	4	3	4	4	**3.78**
Thermal	4	4	4	4	4	4	2	4	4	**3.78**
Ultra Violet	3	3	3	3	3	3	3	3	3	3.00
Visual	2	4	4	4	4	4	3	5	5	**3.89**
Wavelength	3	4	4	4	4	3	3	3	3	3.44
x-Ray	3	3	3	3	4	3	3	3	3	3.11
Avg	3.2272727	3.5	3.5	3.5909091	**3.77272727**	3.3181818	3.22727273	3.2272727	3.272727	3.40

Future Signature Detection Scale

5	Very Difficult
4	Difficult
3	Moderate to Routine
2	Easy to Detect
1	Extremly Easy to Detect (Signatures Avail via Commercial means)

Russia Noise Reduction 1 | Russia Noise Reduction2 Top C&S

Figure 26: Russia Estimated Signature and Threat Capability Matrix Focused on Top Capability and Signatures, Noise Reduction 1

i. Russia will likely focus its modernization across 40% of its main capabilities (8/19). The future threat capabilities that will be difficult to detect are EW (16/22), Tanks (6/22), Air Defense (7/22) and Missile (long- and short-range strike), (8/22) capabilities by leveraging Cyber (12/22), C4ISR (12/22), Deception (13/22) and Stealth (4/22) enabling technologies. Over 68% of these capabilities will likely be difficult to detect in the future as 90 out of 176 signatures in these categories, over 50% become harder, on average, to detect in the 2030-2040 time frame across six primary signatures—including non-technical signatures such as doctrine, electro-optical signatures, and signatures associated with visual, speed, thermal and software changes.

j. Based on a report by the Republic of Estonia Ministry of Defense despite the pace of probable modernization, the Russian armed forces will likely largely rely on a mix of legacy hardware and modernized Soviet systems alongside new designs. It is likely sustained investment in modernization efforts and military R&D is necessary for Russia to remain a credible military power in 2030-2040.

2. Electronic warfare capabilities are likely to see the most upgrades, and consequently, the most negative overall change in their signature profile. In fact, in 16 of the 22 signatures examined, over 72% of Russia's electronic warfare capabilities are likely to get more difficult or much more difficult to detect over the next 20 years.

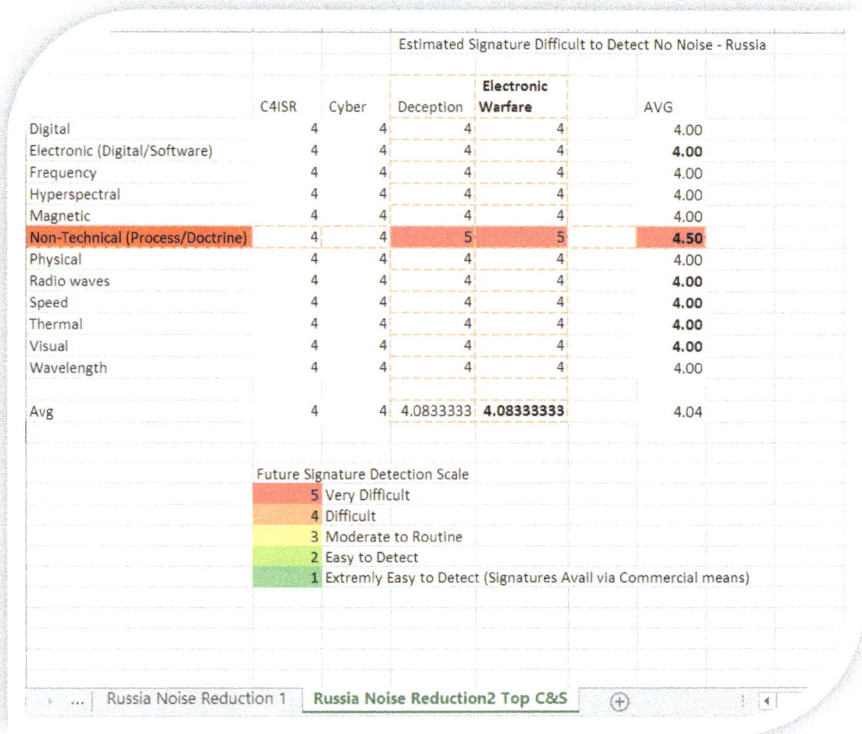

Estimated Signature Difficult to Detect No Noise - Russia

	C4ISR	Cyber	Deception	Electronic Warfare	AVG
Digital	4	4	4	4	4.00
Electronic (Digital/Software)	4	4	4	4	**4.00**
Frequency	4	4	4	4	4.00
Hyperspectral	4	4	4	4	4.00
Magnetic	4	4	4	4	4.00
Non-Technical (Process/Doctrine)	4	4	5	5	**4.50**
Physical	4	4	4	4	4.00
Radio waves	4	4	4	4	**4.00**
Speed	4	4	4	4	**4.00**
Thermal	4	4	4	4	**4.00**
Visual	4	4	4	4	**4.00**
Wavelength	4	4	4	4	4.00
Avg	4	4	4.0833333	**4.08333333**	4.04

Future Signature Detection Scale

5	Very Difficult
4	Difficult
3	Moderate to Routine
2	Easy to Detect
1	Extremly Easy to Detect (Signatures Avail via Commercial means)

Figure 27: Russia Estimated Signature and Threat Capability Matrix Focused on Top Capability and Signatures, Noise Reduction 2

These upgrades will likely impact not only the three signatures named above but also a wide variety of other signatures including those such as thermal, acoustic and hyperspectral. Many of the Russian EW systems are highly mobile, including small systems deployable by UAVs, making targeting and neutralizing them more complex and challenging. These changes are largely due to a result of reforms and modernization efforts. This effort is complemented by changes to organization, doctrine, command structure, training and tactics, as well as techniques and procedures (non-technical signatures).

a. Based on a report by the Republic of Estonia Ministry of Defense Russia has consistently invested in EW modernization since 2009, with modernized EW systems entering service across strategic, operational and tactical levels to augment capabilities of all service branches. Russia has tested close to 200 different military technologies, several EW platforms and capabilities such as the Leer-3 EW platform — a UAV-carried, cell tower-suppressor delivered via a team of two to three Orlan UAVs.

b. Based on reports published by Brookings Russia is increasing automation of EW systems; modern AI algorithms are being investigated to determine their value as a component of new EW systems.

c. According to Roger McDermott from the Estonia Ministry of Defense, 1) a uniting theme among the expert Russian military community is the extent to which they see future synergy between EW and network-centric warfare capability, 2) Russian Ground Forces do not move or conduct operations without EW support, 3) There appears to be a close link between SIGINT, air defense, artillery and EW, which is evident in Russia's application of hard power in south-eastern Ukraine, and 4) Russia's military theorists recognize the EMS as another legitimate domain of warfare, 5) By 2025 or later, the EW Forces could emerge as a new combat arm with a pivotal role in military operations this would mean it moved from a support role to a fully-fledged combat arm.

d. Despite the potential EW technologies advances in reducing EW signatures and improving detection, targeting and deception in unmanned, cyber and missile systems, significant improvement in AI-enabled EW from Russia is unlikely to provide an overwhelming advantage in the electromagnetic spectrum.

3. Russia's Deception, Cyber and C4ISR capabilities are also likely to see significant changes in their signature profiles making them more difficult to detect over the next 20 years. In 37 areas out of 66, over 56% present signature detection challenges, mostly in non-technical, thermal, frequency, software and digital signature profiles. These changes are largely due to Russia focusing its modernization on four critical capabilities: EW, Tanks, Air Defense and Missiles through a complex mix of design features, performance, mission planning, and tactics leveraging Cyber, C4ISR, Deception and Stealth technologies to reduce signatures and emissions from the probability of detection.

   a. Based on reports published by Talal Husseini from army technology.com due to the emergence of new threat signature absorbing and camouflage materials, Russia has placed increasing importance on producing these materials to reduce signatures and emissions from the probability of detection.

   b. According to Timothy Thomas a EUCOM Information Operations Domain Specialist Russia has been testing and evaluating a variety of Electronic Warfare (EW) technologies that can be combined with stealth UAVs which can enter enemy airspace covertly and conduct EW operations. In addition, Russia is modernizing its C4ISR, EW and Cyber platforms to enhance its Air Defense and Missile threat signatures.

c. According to Russian military experts from published Rand reports Russia is seeking to advance and export some of its most capable air defense and missile protective measures to achieve an effective layered air defense. Due to the planned modernization, the Russian Armed Forces will likely increase its strike force and firepower capability as well as various air-defense assets and high-precision systems. Moreover, Russian long-range air defense platforms will likely be coupled with an enhanced upgrade and potential replacement the S-500. Russia will advance and integrate its C4ISR, EW, Cyber and Deception capabilities and tactics to capitalize on its layered air defense initiatives as the S-300, S-400 and future S-500 are most effective as part of a much wider integrated air defense system (IADS).

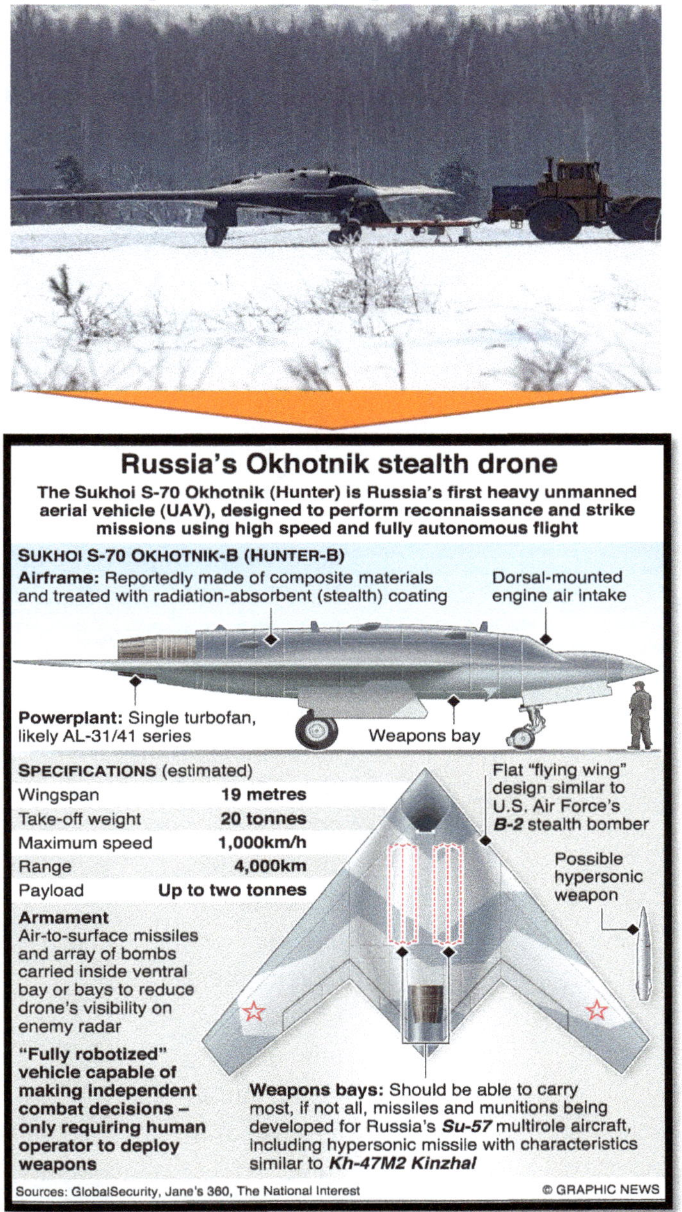

Figure 28: Russia Stealth Drone, the S-70 Okhotnik. As seen above in the photo reportedly Russia's Okhotnik-B stealth attack drone picture leaked. The photo depicts a tractor towing the apparently roughly 50-foot-wide unmanned aerial vehicle along a snow-ringed runway at an airfield in Novosibirsk in southern Russia. Source: Janes, Task and Purpose 2019. Can View at: https://taskandpurpose.com/military-tech/photos-of-russias-new-stealth-drone-just-leaked-heres-what-you-should-know

d. According to the Heritage Foundation despite Russia's air defense and missile sophistication, due to the S-400 currently dependent on a single engagement radar, it has a limited number of firing platforms presenting targeting vulnerabilities. It is likely Russia will not be able to mitigate this vulnerability until it can mass produce the sophisticated S-500 until 2035-2040.

4. Russia's Stealth, Tank, C4ISR, Cyber, EW and Hypersonic visual signatures will likely be difficult to detect due to the emergence of new threat signature absorbing

and camouflage material technologies. Out of the signatures examined Russia's stealth and tank signatures will be more difficult to detect in 10 areas and C4ISR and Cyber more difficult to detect in 12 areas. It will be significantly difficult to detect Russian Tank signatures in 2030-2040 due to reduced visual, thermal, infrared and electronic signatures. These efforts are complemented by the below factors:

a. According to analysts at the Hoover Institution at Stanford University it is highly likely that in 2030-2040 Russia will incorporate many different Air Defense, Missile, EW, Cyber, and C4ISR radars and other sensors and systems for the detection of different types of enemy targets and use Radar Absorbent Materials (RAM) and other adaptive camouflage, or active camouflage technology to reduce and conceal its physical, thermal, frequency, infrared and electronic threat signatures. Russian tank signatures in 2030-2040 will be lighter, smaller, quieter, and faster. Due to increased Russian tank capability and reduced armored signatures, it is highly likely Russia will apply dual use technological advances through the developments in AI and EW to deceive and improve its armored tank weapon capability.

b. Based on published Rand reports and defense and security analysts such as Kyle Mizokami Russia will likely use current and planned tank advancements to make tanks more lethal and smaller as demonstrated in the T-14 with explosive reactive armor, electronic components, enhanced navigation systems, sighting systems, advanced radio and C2 systems, advanced anti-aircraft missiles, anti-aircraft guns, autonomous, IR cooling systems, sensors positioned away from the surfaces and the development of an unmanned turret, and the absence of a fume extractor will likely make these tanks less visible to detect.

Figure 29: Infrared and Invisibility. T-14 Armata's thermal image in the lens of the infrared camera is extremely dim-the only element of the vehicle currently that shows up brightly is the small opening of the engine's exhaust system. Source: Russia Beyond, Maxim Blinov/RIA Novosti

5. Russia's thermal, speed, radio wave, electronic and non-technical (tactics/doctrine) threat signatures from its EW, Deception, Air Defense, Cyber and Hypersonic systems will likely be difficult to detect. Out of the 19 capabilities examined, these signatures profiles represent 40%

(8/19) on average the common signatures that will likely be difficult to detect across the majority of Russia's critical capabilities.

a. Based on published reports from Rand, the Center for Strategic and Budgetary Assessments and this examination multifunctional sensors and systems that can detect, recognize and analyze variations in doctrine, physical, thermal, frequency, infrared, deception and electronic threat signatures and emissions will be required to target Russian Air Defense, Missile, Electronic Warfare (EW), Cyber, and Command, Control, Communications, Computers (C4) Intelligence, Surveillance and Reconnaissance (ISR) (C4ISR) capabilities.
b. Due to advances in missile, radar and stealth technology Russia is highly likely to reduce its air defense and missile physical, thermal, frequency, infrared and electronic signatures through a complex mix of design features, performance, mission planning, and tactics.

6. To analyze and target the critical Russian EW, Deception, Cyber, C4ISR, Tank, Hypersonic, Air Defense, Missile and Stealth capabilities in 2030-2040, multifunctional sensors and systems that can detect and recognize, electronic/digital, visual, non-technical, speed, physical, frequency, and magnetic signatures will likely be required. These capabilities (9/19) and signature profiles (7/22) account for 70% of the likely future threat signatures that will require sensors and systems to recognize, detect, analyze and target. The U.S. Army will likely need Measurement and Signature Intelligence (MASINT) sensors and systems that can detect, track, identify or describe the distinctive characteristics of Russian fixed or dynamic visual, thermal, radio waves, speed, digital and non-technical signatures.
    a. Based on reports published by Talal Husseini from army technology.com to avoid detection from high-end thermal imaging and MASINT, Russia has placed increasing importance on developing threat signature absorbing, camouflage and deception materials.
    b. According to Brookings published reports and Talal Husseini and data from this report due to increased Russian tank capability and reduced armored signatures, Russia will likely significantly decrease its visual, thermal, electronic, digital and radio wave threat emissions and signatures to conceal its speed, composition and intentions.

Likely At Least 80% of Russian Military Modernized by 2030

Executive Summary:

Despite Russia's defense budget likely reduced to approximately 3% of GDP on defense purposes in the next 5-10 years, based on Rand published reports Russia is likely to improve and develop new advanced weapons systems such as long-range strike systems, C4ISR, EW, Unmanned systems and air defense capabilities—approximately 80% of its force-- to strengthen strategic deterrence by 2030.

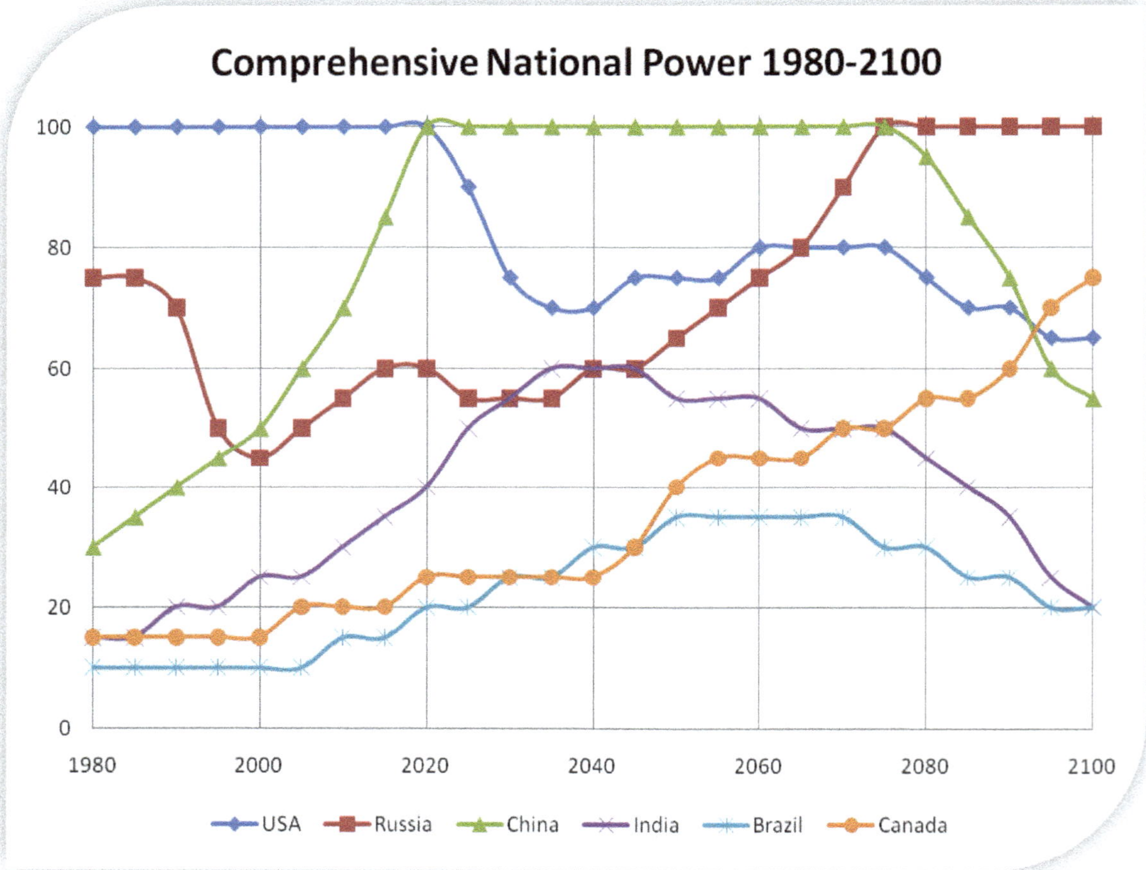

Figure 30 Future Global Power Trends as Comprehensive National Power as a base (an index of power that attempts to express a nation's economic, military, and cultural power in a single number), that stress energy resources, human capital, economic growth, and climate change factors in Anatoly Karlin's analysis of Great Power Predictions. [1] Source: Anatoly Karlin, Future Superpowers – The World To 2100 https://akarlin.com/2011/06/future-superpowers/.

Discussion:

Despite Western sanctions, Russia laid out in strategic security documents the focus of its military modernization: To accelerate the upgrading of weapons and equipment and to commission new weapons in order to achieve the minimum goal of 70% modernized equipment to the Russia military by 2020. [M] It is likely the majority of modernization will achieve the stated goal in 2020 and Russia will focus its future modernization on four critical capabilities: EW, Tanks, Air Defense and Missiles through a complex mix of design features, performance, mission planning, and tactics leveraging Cyber, C4ISR, Deception and Stealth technologies to reduce signatures and emissions from the probability of detection. Figure 30 predicts future global power trends and suggests that Russia will decline in power from 2030-2040 but will increase and continue to rise starting in 2040. [M]

As illustrated in Figure 30, Russia faced a difficult fiscal situation in 2015 and 2016, including declines in economic growth. According to Defense Policies of Countries reports, Russia has implemented full-scale military reform since 1997 by presenting the three pillars of reform: downsizing; modernization; and professionalization. Furthermore, Russia has worked to secure a budget for national defense in order to achieve the goal of modernizing its equipment by 20230 and other targets. Regarding the modernization of the military forces, Russia is working to increase its percentage of new equipment. It highly likely that Russia will continue modernization efforts in the future. [H]

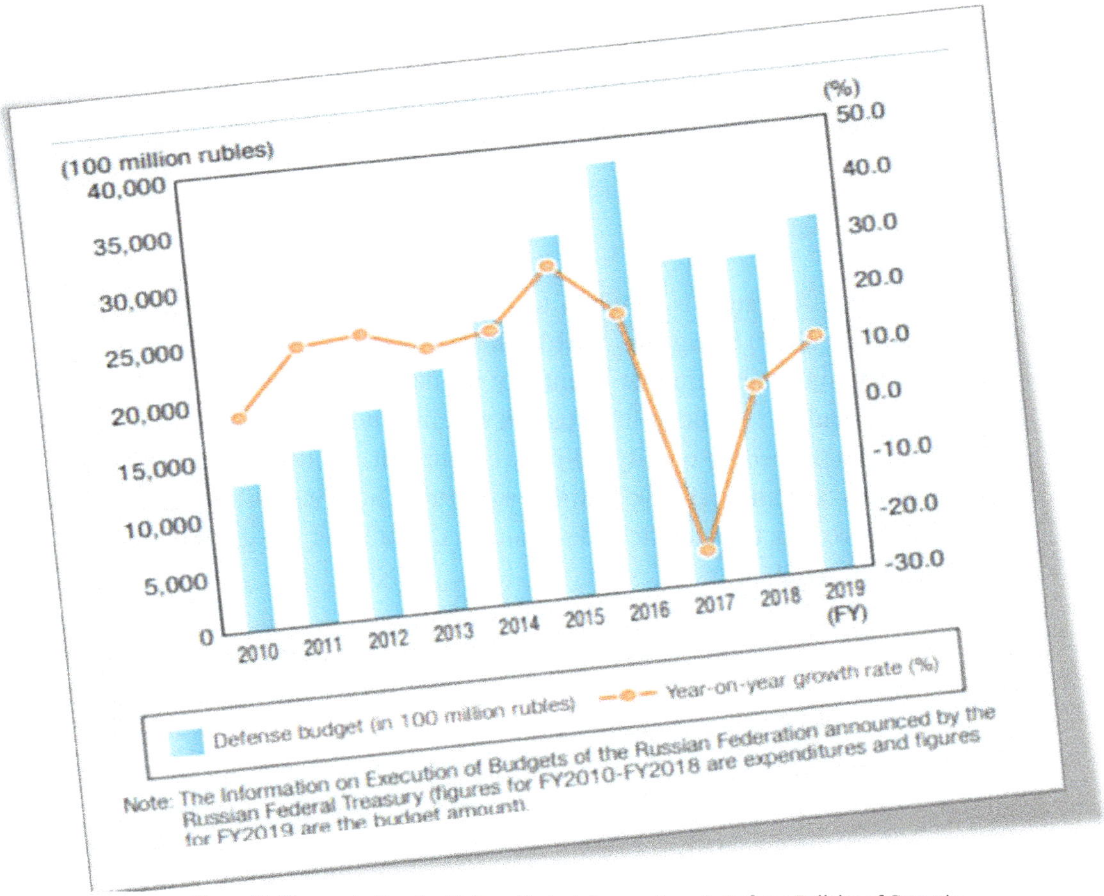

Figure 31 Changes in Russi'as Defense Budget. Source: Japan's Defense Policies of Countries. https://www.mod.go.jp/e/publ/w_paper/pdf/2019/DOJ2019_1-2-4.pdf

During Russian President Putin's address to the Collegium of the Military-Industrial Commission on December 22, 2017 at the newly built training facility of the Academy of the Strategic Rocket Forces in the city of Balashikha outside Moscow, set the objective of

Type of weapons system	2013	2015	2017	2020
Submarines	47	51	59	71
Surface ships	41	44	54	71
Aircraft	23	37	55	71
Helicopters	39	63	76	85
Ground missile systems	27	64	100	100
Artillery	51	53	59	79
Armoured vehicles	20	37	56	82
Multi-role vehicles	40	48	56	72

Source: Ministry of Defence of the Russian Federation (undated), 'Plan deyatel'nosti na 2013–2020 gg' [Action Plan for 2013–2020], Moscow: Ministry of Defence, http://mil.ru/mod_activity_plan/constr/vvst/plan.htm (accessed 21 Feb. 2018).

Figure 32 Reported share of modern military equipment in total inventories for selected categories of weapons system, per cent (2013-17 = actual; 2020 = target) Source: Chatham House The Royal Institute of International Affairs.[M]

harnessing the potential of the defense industry to manufacture modern, high-tech, and competitive civilian products. [L]

Due to the State Armaments Program to 2027 Gosudarstvennaya Programma Vooruzheniya, (GPV) being fine-tuned based on lessons from the conflicts in Ukraine and Syria, Russia is signaling its intent that military modernization is progressing in quantity and quality. [M] Due to the Russian defense industry being overwhelmed when it comes to mass-producing quality military equipment, achieving the twin goals of high-volume production and the development of entirely new weapon systems as illustrated in Figure 32 is proving to be difficult to achieve, threatening the goal of 70% of full military modernization by 2020. [M]

The chief of staff of the Russian armed forces, Valeriy Gerasimov set a priority of strengthening command-and-control (C2) systems for the armed forces – including, notably, intelligence, surveillance and reconnaissance (ISR) capabilities to achieve and increase network-centric warfare capabilities. The ground forces are expected to obtain a fully operational tactical automated C2 system with modern ISR and electronic warfare (EW) components for artillery. Figure 33 video illustrates a recent 2018 Russia miliary exercise testing combat readiness with its artillery, unmanned, air, ground, and sea platforms. This is in line with GPV priorities concerning force mobility, force deployability and strengthened C2. Tank units are likely going to be modernized, with T-72B3s,T-80BVs, T90's and T14's. [M]

According to Russian military analyst Michael Kofman and political scientist Maxim Trudolyubov Russia has gone through transformative military reforms and modernization and Russia's yearly defense budget is somewhere between $150 and $180 billion and the state of Russia's military reforms and modernization programs is focused on conducting long-term investment in defense modernization and defense planning to advance its military capabilities. The United States spends about 30% of its defense budget on procurement and R&D, while Russia spends close to 50%.[M] Russia is likely to continue to challenge U.S. policy abroad, gain leverage, and raise the transaction costs by launching additional destructive campaigns through indirect warfare in 2030-2040 as an indirect forms of competition requires less money and resources to sustain it while spreading thin your adversary's forces and cause it to diffuse its resources.

Figure 33 Vostok 2018, Largest Russia Military Exercise. Click on picture or go to: https://youtu.be/zc-fyC-z29U to view video. Source: Euronews

Different organizations within the Russian policy-making community have developed competing strategies for the country's future economic development given the aims in the 2027 State Armaments Program. These organizations include the Centre for Strategic Research (CSR) and the Stolypin Club, a collection of business representatives and policy officials with an interest in stimulating a faster rate of economic growth. Their respective strategies have been presented to the country's leadership in the hope that these might form the basis of economic policy between 2018 and 2024. Each organization has drawn up projections of Russia's future economic performance, based on different assumptions about the direction of economic policy, see Figures 3 and 4. According to the projections developed by the CSR and the Stolypin Club, Russia is likely in a position to at least come close to allocating the required R19 trillion on military equipment. If GDP growth rates exceed the more pessimistic projections, it is possible that GPV 2027 might be fully funded without too much difficulty. [M]

Due to the planned modernization, the Russian Armed Forces will likely increase its strike force and firepower capability as well as various air-defense assets and high-precision systems. [M] Russia will continue to improve and develop new advanced weapons systems such as long-range strike systems, C4ISR, EW, Unmanned systems and air defense capabilities. [H] Due to the positive economic growth in 2018, it is likely Russia will come close to achieving its military modernization goal, but will not achieve 70% military modernization in 2020. [M]

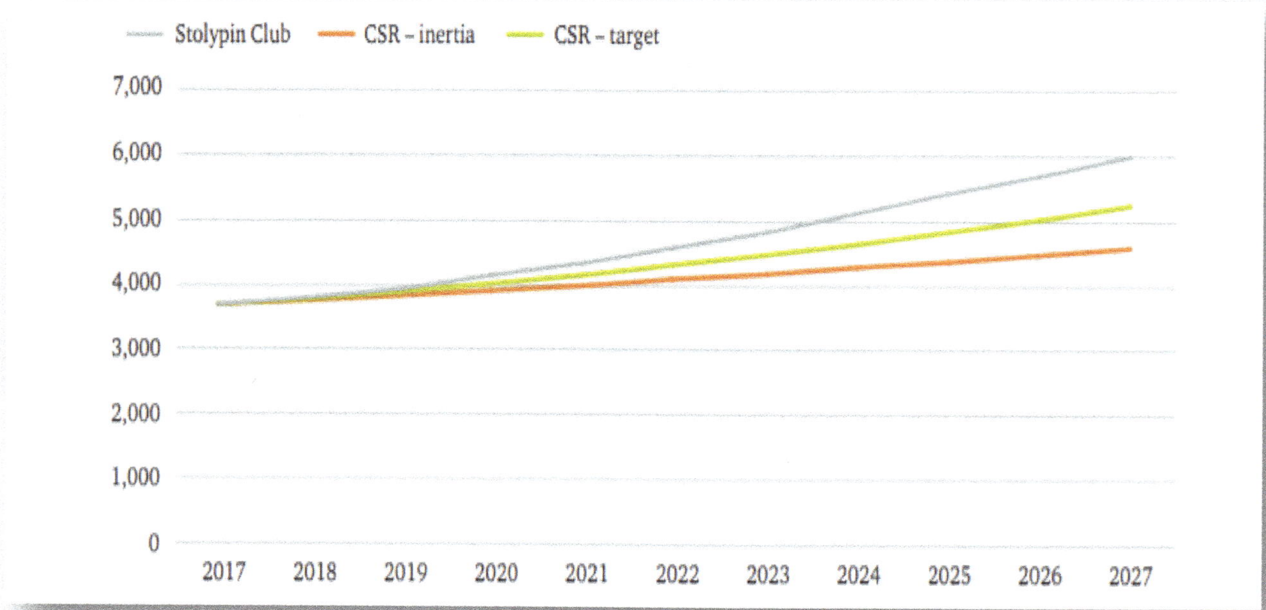

Figure 34 Rouble value of total defense expenditures projections 2017-2027. Source: Stolypin Club (2017), Center for Strategic Research (2017).

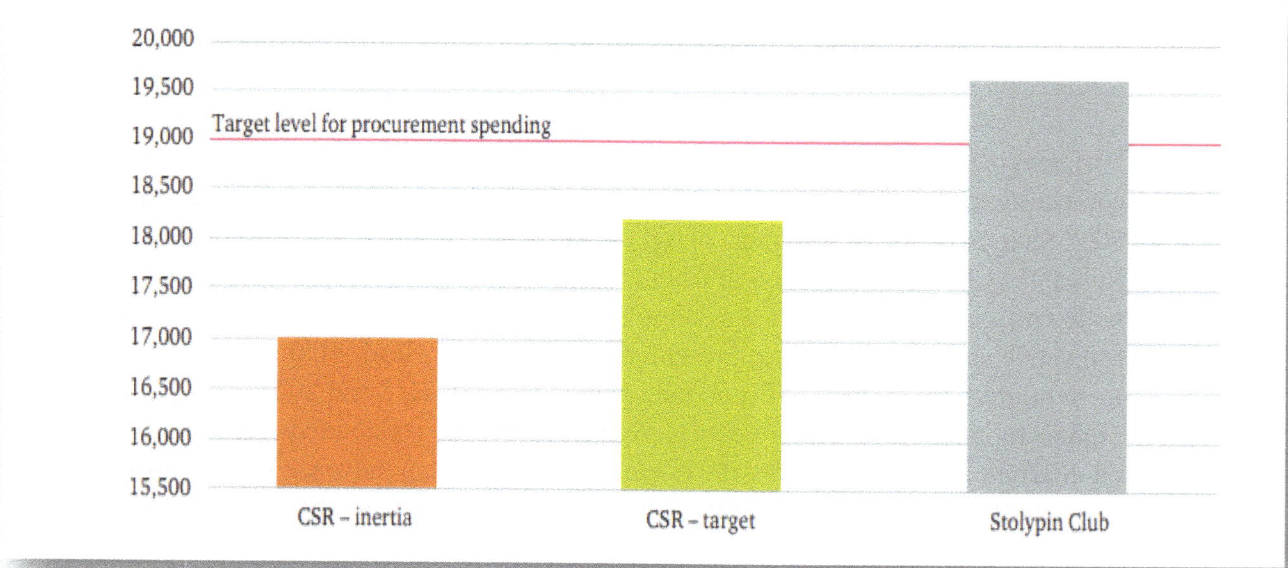

Figure 35 Cumulative spending on the State defense order under GPV 2027 alternative growth projections, 2018-2027. Source: Stolypin Club (2017), Center for Strategic Research (2017).

Despite the Russia armament program being classified, adequate details have been published by Russian officials, politicians, and press articles, which permit a rough estimate of the further development of armed forces until at least 2030. [M] According to The State Armaments Programme for 2018-2027, Russia will invest 19 trillion rubles (about 306 billion US dollars) in the next 10 years in defense procurement and equipment

upgrades. [M] Specifically, Russia is planning to develop and procure high-precision weapons for air, sea and land battle – including hypersonic missiles – unmanned air strike complexes, individual equipment for servicemen and advanced reconnaissance, communication and electronic warfare systems in the next 5-10 years. [M]

Russia is likely to continue its advances in its air defense program and will likely threaten the reach of the U.S. military. [M] As illustrated in Figure 35 Russian air defense doctrine follows a three-tier approach, a layered system that allows Russian air defense forces to create anti-access area-denial (A2AD) zones that can be difficult to penetrate. The highest tier of these defensive networks uses long-range systems such as the S-200 and S-400, providing air defense bubbles potentially up to 800 km in diameter. [H]

Due to the hybrid warfare, cyber warfare and information warfare lessons learned from 2014 to date in Ukraine and in Syria by Russia military operations, new requirements for Russian equipment development have emerged. Russian President Putin has clearly pointed out, "Russia must put the development of digital technology equipment, artificial intelligence, drones and robot systems on its agenda in terms of Russian military construction." [M]

Due to recent Russian military exercises as illustrated in Figure 33 video that were conducted in 2020, Russia is likely to modernize and integrate its unmanned vehicle programs to assist targeting through leveraging artillery to increase their long-range accuracy for increased lethality. [H] During this exercise involving 2,000 troops and 400 pieces of equipment, the Orlan-10 UAV was used to reconnoiter potential targets and provide real-time co-ordinates of high-priority targets for 2S5-equipped units. [H]

Despite global power trends illustrated in Figure 30, it more likely by 2030, the Russian armed forces are likely to be considerably better equipped than they are today. [M] Due to the pace of probable modernization, measured progress likely in the development of new-generation equipment, and the Russian armed forces will likely still rely on a mix of legacy hardware and modernized Soviet systems alongside new designs. [M] It is likely sustained investment in modernization efforts and military R&D is necessary for Russia to remain a credible military power for at least the next 10-15 years. [M]

Figure 36 Russia's Air-Defense System, Click on picture to enlarge or go to: https://i1.wp.com/www.offiziere.ch/wp-content/uploads-001/2019/01/zapad-001.png?ssl=1 to view. Source: Offiziere.ch Security Policy – Armed Forces – Media

Analytic Confidence:

The analytic confidence for this estimate is *moderate*. Open sources were used, source content varied; however, sources tended to corroborate one another and were generally reliable.

Authored by: Sam Smith

Russian AI-driven EW Capability in 2030-2040 Highly Likely to Use High Power Signals Emitted Throughout A Large Range of Frequencies to Overload the Receivers of Enemy Electronic Systems

Executive Summary:

Due to advances in digital technology have made it possible for Russia and modern militaries to develop highly flexible and adaptable electronic systems with feedback that enable rapid adaptation to the electromagnetic environment. Russia will highly likely use high power signals in 2030-2040 emitted throughout a large range of frequencies to overload the receivers of enemy electronic systems and as a result will develop material to shield electronics from EW countermeasures and field fabric to reduce radar signatures. Despite Russian enhancements in detection, targeting and deception expertise with advances in AI-driven EW through unmanned, cyber and missile systems, it is unlikely to be a sudden, significant improvement in AI-enabled EW from Russia that would provide an overwhelming advantage to Russia in the electromagnetic spectrum.

Figure 37: Drone based Jamming. The Leer-3 Command Vehicle and Orlan-10 type EW drone. Source: Vitaly Kuzman. Can view at: http://dl.icdst.org/pdfs/files3/906f2544ddd693eb1118881a5baff0a3.pdf

Discussion:

Russia is willing to introduce entirely new ways of using EW assets, Russia will highly likely use high power signals emitted throughout a large range of frequencies to overload the receivers of enemy electronic systems. Despite this capability, it presents detection and targeting opportunities as this high-power signal requires a lot of energy that will cause increased electrical, digital, infrared, radiation, and thermal signatures in the vicinity of the EW stations. Due to this energy (signature) vulnerability it is likely in 2030-2040 that Russia will use advances in AI technology to automate its EW action resulting in reduced EW signatures. Due to these advances, Russia will likely inject false location data to users of GPS for navigation and timing targeting and likely will present some difficulty in detection and recognition. With the increasing automation of EW systems, Russia is investigating modern AI algorithms to determine their value as a component of new EW systems. [H]

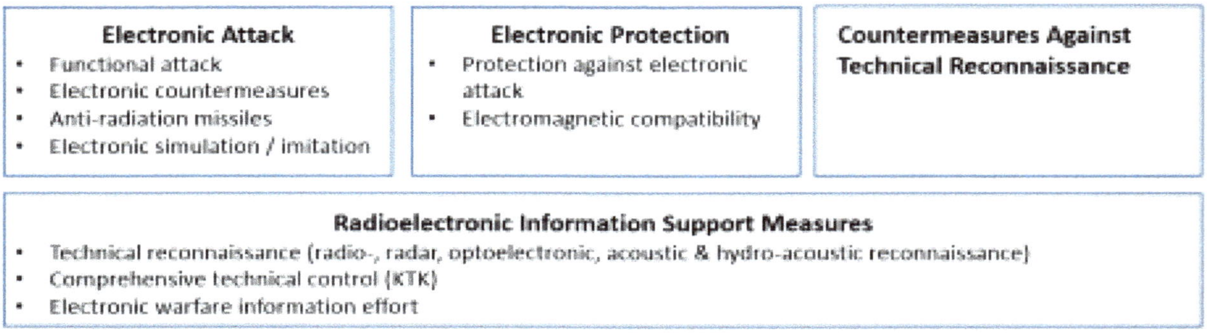

Figure 38: The Modern Russia EW definition with its four subdivisions. Source: Guzenko & Moraresku, can view at: http://dl.icdst.org/pdfs/files3/906f2544ddd693eb1118881a5baff0a3.pdf. .

According to Russian data, the input made by EW units in the missions such as de-organizing the enemy's C4ISR and artillery control systems may be higher than 70%. Due to this a range of new, multi-purpose systems is being intensely introduced into the Russian military inventory, designed to "attack the enemy in every area (in space, in the air, within land and sea), and within the whole depth of his positions", along with new EW systems. [H] These new systems designed according to completely new principles and with the use of new technologies, along with the traditional systems, are to increase the efficiency of its Armed Forces. Furthermore, according to Russia news media outlets when it comes to military applications of artificial intelligence, Russia's investments in artificial intelligence and other emerging technologies will likely advance in 2030 to counter the physical, cognitive, and operational challenges of likely future urban warfare. [L] Due to Russia likely improving its AI-driven military application capability Russia's Advanced Research Foundation announced it will design an AI-enabled system to sift through vast quantities of satellite imagery for faster and more accurate analysis. [M] This demonstrates how the Russian defense community is thinking through the integration of different systems in combat and how they can apply this development to other capabilities such as EW platforms.

Due to Russia likely improving its AI-driven EW capability, detection, targeting and deception expertise will advance in their unmanned, cyber, and missile systems. In 2030-2040 it is highly likely a highly contested EW environment will increase in complexity and due to human control over unmanned platforms, sensors, and, in particular, weapons may become increasingly unreliable. Despite the potential EW technological advances there is unlikely to be a sudden, significant improvement in AI-enabled EW from Russia that would provide an overwhelming advantage to Russia in the electromagnetic spectrum outside its limited advances in reducing its EW signatures and improving detection, targeting and deception in unmanned, cyber and missile systems. [H]

EW technology is inherently automated due to the rapid speed of signal generation, propagation, and processing. Advances in digital technology have made it possible for Russia and modern militaries to develop highly flexible and adaptable electronic systems with feedback that enable rapid adaptation to the electromagnetic environment. [H]

Another field of research that Russia is likely to improve is the development of ferrite fibres. This material can shield electronics from EW countermeasures and create a fabric to reduce radar signatures. The use of ferrite fibres was exemplified in connection with the S-500 missile system and the Armata T-14 battle tank. Procurement of these systems has not yet begun, and it is likely in 2030-2040 that ferrite fibre will be used for electromagnetic hardening and signature reduction in these two weapon systems. [M]

The introduction of unmanned EW systems as a mean to advance an EW combat operation on to enemy territory is a stated Russia objective and will likely advance this capability by 2030. The use of UAVs in EW is likely to significantly increase as the use of quadcopters as antennae for ground-based EW systems allows the use of a non-aerodynamic antenna, thereby opening up a wide range of different configurations for jamming and direction finding with higher payloads.

It is highly likely that Russia will use high power signals emitted throughout a large range of frequencies to overload the receivers of enemy electronic systems. Despite this capability, it presents detection and targeting opportunities as this high-power signal requires a lot of energy that will cause increased electrical, digital, infrared, radiation, and thermal signatures in the vicinity of the EW station. [M] Russia is highly likely to disrupt communication signals that operate within the frequency band 3 to 25MHz. Its purpose is to jam the short-wave transmission at the distances of up to 5000 kilometres, making this solution one of the most powerful ECM suites when it comes to the signal power, exceeding 400 kW. Despite this capability, the whole system is based on seven Kamaz trucks, and it is not a mobile package and requires 72 hours of daylight required to deploy the antenna array to accomplish effective communication disruption. [M] Despite the criticality to suppress enemy command and control systems, enhancing the survivability of military forces and protecting critically important infrastructure through EW means is paramount to Russia. [H]

The Russian Armed Forces rely heavily on the ability of domestic industry to provide them with modern EW equipment. More than 120 Russian companies were involved in the development and production of EW systems and around ten of these companies are responsible for the majority of EW equipment output. Figure 38 shows the location of 13 of the most important EW companies.

Russia has established a few innovation centers to drastically lead in AI. Furthermore, there has not been an enduring emphasis from Russian leadership for taking a leading role in modern AI algorithm development, as compared with China's repeated emphasis over the past few years. It is more likely Russia will continue to make progress in improving the responsiveness and speed of their EW systems. [H] Autonomous and independent EW companies have also become a part of all the motorized and armored brigades and land forces divisions, as well as of all brigades of the airborne [VDV] units. Whereas the Russians have assumed that the VDV units, as the first component, until the year 2020, would receive brand new EW equipment. [M] The officers of the EW units are being trained by the Military Aviation Engineering University (Voronezh). This University also has an R&D institute for EW embedded within its structures. Junior EW specialists, who are to work within all branches of the Russian Armed Forces, are being prepped at the general military training center in Tambov. [M] Almost all EW equipment for the Russian Army is currently being delivered by the Radio-Electronic Technologies JSC (KRET) plant. The umbrella company mentioned above, throughout the period between 2009 and 2012, has fused most of the Russian facilities dealing with supplies of military radio-electronics. [M]

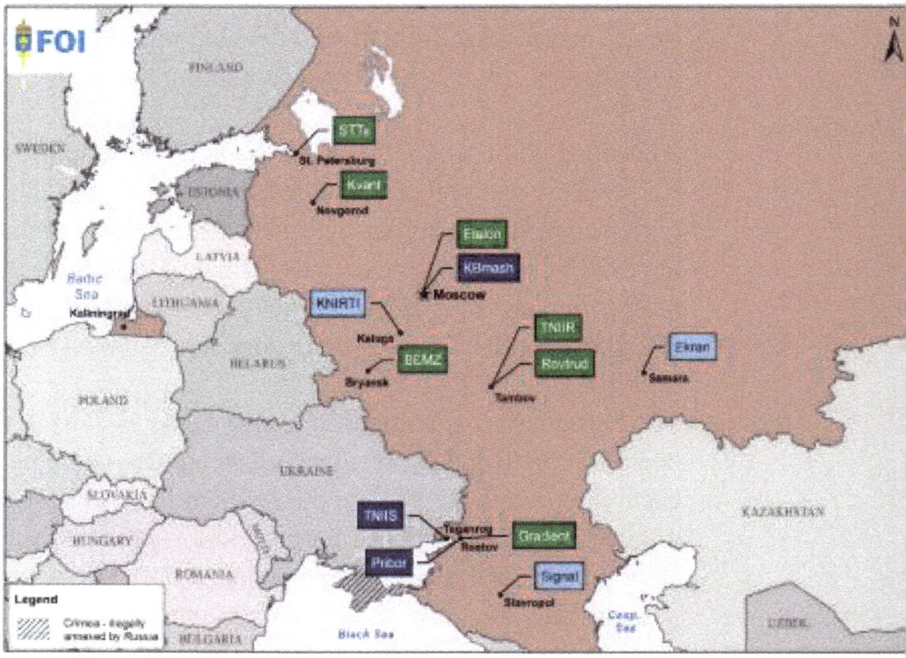

Figure 39: Key KRET and Sozvezdie EW companies and their location, by branch of service. Legend: naval equipment = dark blue; aerial equipment = light blue; ground-based equipment= green.Source: FOI, can be viewed at http://dl.icdst.org/pdfs/files3/906f2544ddd693eb1118881a5baff0a3.pdf

Analytic Confidence:

The analytic confidence in this estimate is moderate. The reliability of the sources available on this topic were above average with several high-quality sources available for the estimate. The sources available did tend to corroborate each other and analyst collaboration was very strong. The estimates contained within the sources are good, but given the length of time frame of the estimate, this report is sensitive to changes in market and economic conditions

.

Authored by: Sam Smith

Russian is Highly Likely to Significantly Reduce Electronic, Acoustic, Infrared, Magnetic, Physical, Thermal and Seismic Signatures on Tanks in 2030 – 2040 Leveraging Stealth, EW, and AI Technologies

Executive Summary:

Russia's tanks are becoming more modernized, capable, and will be harder to detect than previous versions. Despite Russia's defense budget likely reduced to approximately 3% of GDP on defense purposes in the next 5-10 years, due to leveraging stealth, EW, and AI technological advancements, Russia will likely revolutionize its Main Battle Tank (MBT) program by 2040 and is highly likely to significantly reduce acoustic, radar, and infrared signatures presenting challenges to detect, recognize and target.

Discussion:

It is highly likely that future Russian tanks will significantly reduce their infrared, acoustic, and radar signatures which would indicate that in 2030-2040 Russian tanks will be able to blend in with their heat surroundings better, due to leveraging stealth technology that reduces or masks the temperature of their exhausts and the heat from weapon barrels. [H]

Figure 40: Infrared and Invisibility. T-14 Armata's thermal image in the lens of the infrared camera is extremely dim-the only element of the vehicle currently that shows up brightly is the small opening of the engine's exhaust system. Source: Russia Beyond, Maxim Blinov/RIA Novosti: [H]

Based on published Rand reports and defense and security analysts such as Kyle Mizokami, due to advanced materials, innovative automotive systems, autonomous technology, new weapon systems and active protection systems, it is likely that Russian tank signatures in 2030-2040 will be lighter, smaller, quieter, and faster. [M] Moreover, it is likely that Russian tanks by 2035 will have a reduced electronic, acoustic, infrared, magnetic, physical, thermal and seismic signature making the tank significantly harder to detect . [M] Due to increased Russian tank capability and reduced armored signatures, it is highly likely Russia will apply dual use technological advances through the developments in AI and EW to deceive and improve its armored tank weapon capability. [M]

Despite future planned advancements, currently, the Russian army mainly deploys a mixed force of T-72s, T-80s and T-90s tanks. The Russian Army currently fields around 40 combat brigades, each with one battalion of tanks. Russian defense minister Sergei Shoigu stated in 2020 that over 400 new and upgraded armored vehicles are scheduled for fielding. The Russian army will receive its first new T-90M tanks in 2020 as part of the 400 armored vehicle fielding plan. [H]

The T-90M, is highly likely to be Russia's most advanced front-line tank for the next 10 years, it is a modernized version of a tank that first entered service in 1993. The Defense Minister also stated that the Russian Tank modernization would include T-72B3M tanks with the domestic sighting and observation system, T-90M Proryv-3 and T-80BVM tanks and BMP-1AM infantry fighting vehicles. [M]

The T-90M features a new engine and the latest version of the Relikt explosive reactive armor kit, this new engine is smaller which creates less heat and acoustic signatures. It is likely that at least a Battalion set of the new modernized tanks, the T-90M, will be fielded to the Russian Army, approximately 40 tanks in 2020. The chances are a little better that at least 100 more will be produced and fielded in the next 10 years.

Despite the more sophisticated, T-90M, T-64/80 family of tanks and the next generation Russian MBT, the T-14 Armata, Russian factories will likely continue to produce the T-27/90 family of tanks over the next 10 years due to them being simpler to build, support and operate than the T-90M and the T-14. However, it is likely that in 2030-2040 Russia will further modernize its T-14, autonomous, and stealth tank capability due to leveraging advancements in technology. [M]

Russia will likely use current and planned tank advancements to make tanks more lethal and smaller as demonstrated in the T-14 with explosive reactive armor, electronic components, enhanced navigation systems, sighting systems, advanced radio and C2 systems, advanced anti-aircraft missiles, anti-aircraft guns, autonomous, IR cooling systems, sensors positioned away from the surfaces and the development of an unmanned turret, and the absence of a fume extractor will likely make these tanks less visible to detect. [M]

Analytic Confidence:

The analytic confidence in this estimate is *moderate*. The reliability of the sources available on this topic were above average with several high-quality sources available for the estimate. The sources available did tend to corroborate each other and analyst collaboration was very strong. The estimates contained within the sources are good, but given the length of time frame of the estimate, this report is sensitive to changes in market and economic conditions.

.

Authored by: Sam Smith

Russian Air Defense & Missile Physical, Thermal, Frequency, Infrared and Electronic Threat Signatures Highly Likely to Decrease in 2030 - 2040

Executive Summary:

Russia in 2030-2040 is highly likely to decrease its air defense and missile physical, thermal, frequency, infrared and electronic signatures through a complex mix of design features, performance, mission planning, and tactics. It is highly likely multifunctional sensors and systems that can detect, recognize and analyze variations in doctrine, physical, thermal, frequency, infrared, deception and electronic threat signatures and emissions will be required to target Russian Air Defense, Missile, Electronic Warfare (EW), Cyber, and Command, Control, Communications, Computers (C4) Intelligence, Surveillance and Reconnaissance (ISR) (C4ISR) capabilities.

Discussion:

Due to advances in missile, radar and stealth technology Russia is highly likely to reduce its air defense and missile physical, thermal, frequency, infrared and electronic signatures through a complex mix of design features, performance, mission planning, and tactics. H

Figure 41: *A battery of deployed S-400 surface-to-air missiles stands in launch configuration in Crimea. Source: SERGEI MALGAVKO/TASS via Getty Images. Can view at:* H

Russia is seeking to advance and export some of its most capable air defense and missile protective measures to achieve an effective layered air defense that include the S-300 and S-400 long-range Surface-to-Air Missile (SAM) systems, that can be ground-to-air missiles or surface-to-air guided weapons. M These Russian long-range air defense platforms will likely be coupled with an enhanced upgrade and potential replacement that is in development. H The S-500 is the system Russia has developed but cannot currently mass produce given the cost and procurement challenges. M It is likely to be massed produced in the next 10-15 years, in the form of new missiles and radars. The S-500 extends the range, and ability to strike at different aircraft, cruise and ballistic missile targets and likely have sophisticated sensors with anti-stealth capability. H Their ranges allow them to target key enemy enabler aircraft, such as valuable aerial refueling tankers and airborne early warning and control aircraft. H Their flexible targeting capabilities mean they can defend against multiple different types of threats and attacks leveraging their EW, Cyber, Deception and C4ISR capabilities and tactics. Despite their currently limited anti-stealth capability, it is likely in 10-15 years

Russia will have the potential to target sophisticated aircraft with their advancements in anti-stealth. [H]

Due to the emergence of new threat signature absorbing and camouflage materials, Russia has placed increasing importance on producing these materials to reduce signatures and emissions from the probability of detection. [M] According to analysts at the Hoover Institution at Stanford University it is highly likely that in 2030-2040 Russia will incorporate many different Air Defense, Missile, EW, Cyber, and C4ISR radars and other sensors and systems for the detection of different types of enemy targets and use Radar Absorbent Materials (RAM) and other adaptive camouflage, or active camouflage technology to reduce and conceal its physical, thermal, frequency, infrared and electronic threat signatures. [M]

Due to the S-400's ability to use four different types of missiles with different weights and capabilities, the system itself forms a large portion of a layered air defense network. This makes the S-400 a more flexible system capable of using missiles employed by earlier S-300 variants. Russia's modernization of its C4ISR, EW and Cyber platforms enhances its Air Defense and Missile threat signatures. They are decreased using multichannel frequencies and electronic signatures to increase the effectiveness of their guidance beams to guide missiles to different targets simultaneously. [H]

Russia has deployed the S-300 and S-400 to operate as stand-alone systems and it is likely that Russia will disperse these air defense and missile forces to further reduce its threat signature effectively limiting the maximum range of its missiles. Despite this limitation it is highly likely that in 2030-2040 Russia will advance and integrate its C4ISR, EW, Cyber and Deception capabilities and tactics to capitalize on its layered air defense initiatives as the S-300, S-400 and future S-500 are most effective as part of a much wider integrated air defense system (IADS). [M]

However, despite these likely advances, unless target data can be provided and updated during the missiles in flight by airborne or forward deployed radars, the full 400km technical range of its missiles cannot be effective against targets below 3000 meters. [L] It is more likely Russia could master this capability closer to 2040. Despite Russia's air defense and missile sophistication, due to the S-400 currently dependent on a single engagement radar, it has a limited number of firing platforms presenting targeting vulnerabilities. It is likely Russia will not be able to mitigate this vulnerability until it can mass produce the sophisticated S-500 until 2035-2040. [M]

Analytic Confidence:

The analytic confidence in this estimate is *moderate*. The reliability of the sources available on this topic were above average with several high-quality sources available for the estimate. The sources available did tend to corroborate each other. The estimates contained within the sources are good, but given the length of time frame of the estimate, this report is sensitive to changes in technology and resources.

Authored by: Sam Smith

Russia: Data Transfer, Networking Content Distro Tech Breakthrough All Likely by 2035

Executive Summary:

Russian platforms in new data transfer, networking, and content distribution technologies are likely to be achieved by 2035, due to Russia's policy prioritizing information and communication technologies (ICT). Russian President Putin stated "During all these years since the unilateral US withdrawal from the ABM Treaty, we have been working intensively on advanced equipment, data transfer, information technology and arms, which allowed us to make a breakthrough in developing new models of strategic weapons". A significant growth of the quantity of data for analysis presents challenges for Russian military decision making. Despite Russia's financial difficulty, its military will likely draw on Russian proficiency in designing advanced weapons as well as these emerging capabilities on today's battlefields.

Discussion:

According to Russian military experts from published Rand and Hudson Institute reports based on Russian data, Information and communication technology (ICT) promotion is one of the Russian policies' priority areas. Hardware continues to dominate the Russian IT market, making up more than 50 percent of the sector and these seven critical technologies are listed in Figure 42. M

- Telecommunication technologies
- Data processing and analysis technologies
- Hardware components, electronic devices and robotics
- Predictive modeling and simulation
- Algorithms and software
- Information Security
- Computer architecture and systems

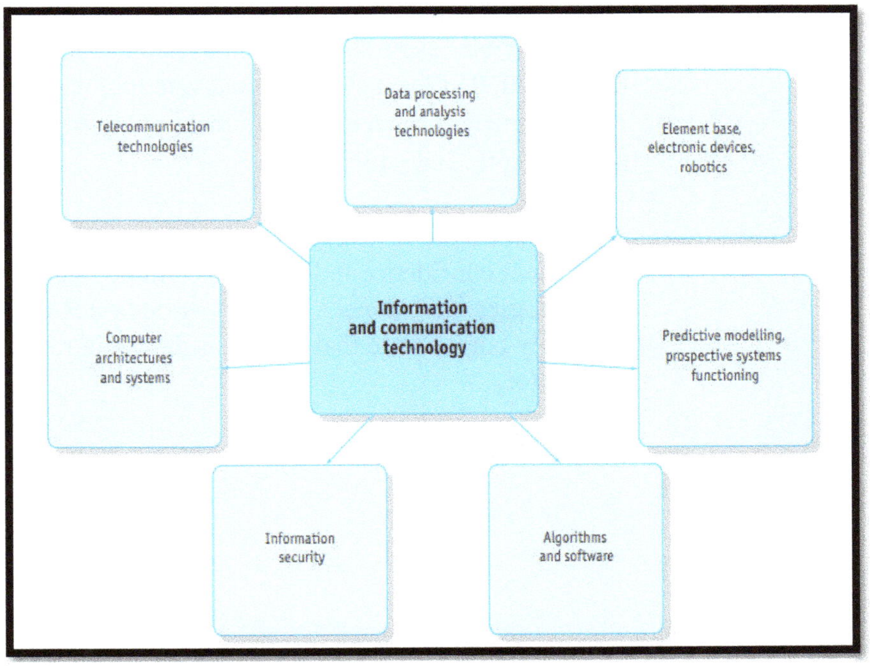

Figure 42: Russian Thematic Fields of the Information & Communication Technology Priority Area

Due to Russia's policy prioritizing information and communication technologies (ICT). Russian President Putin stated "During all these years since the unilateral US withdrawal from the ABM Treaty, we have been working intensively on advanced equipment, data transfer, information technology and arms, which allowed us to make a breakthrough in developing new models of strategic weapons".[M] A significant growth of the quantity of data for analysis presents challenges for Russian military decision making. Despite Russia's financial difficulty, its military will likely draw on Russian proficiency in designing advanced weapons as well as these emerging capabilities on today's battlefields.

Despite Russia's financial difficulties, its Information Technology (IT) market is growing.[H] Russia is surging with its research and development data processing technologies to solve the Big Data problem including personal analytical systems, tools for real-time data processing by 2030.[M] Russia's new data transfer technologies allows terabit data transfer rates. Promising Russian telecommunication technology has the potential for data transfer and interaction between various optical network devices without converting signals into electrical form.[M] Russia is developing new tools and protocols for managing input/output and selection of data, and transparent interaction requests for global data storage systems; wide-area data access, transfer and request tools.[M] According to the machine intelligence laboratory at the Moscow Institute of Physics and Technology, the need for a Russia-wide platform where scientists, researchers, engineers, and entrepreneurs could share their work and cooperate is going to be developed, and said that the country's tech drive urgently needed more engineers, researchers, and computer programmers.[M] Russia's skilled engineers and mathematicians are a valuable resource for tech and defense industry and according to Russian internal data by 2030 at least every second scientist will be under the age of 40.[M] Russia like most countries are weaponizing technology for its military, and they are rapidly advancing this capability.[M]

Analytic Confidence:

Analytical confidence in Russia's technological advancements in these critical technologies is *moderate*. Sources varied on Russia's capability. Despite this, sources did generally corroborate that Russia is applying resources to its research infrastructure in this sector.

Authored by: Sam Smith

Russia Likely to Develop and Begin to Produce A Combat Capable Stealth UAV with EW Capability Within 10 Years That Can Conduct EW Operations at Close Range Within Enemy Airspace

Executive Summary:

Due to rapid advances in mature technology, Russia has been testing and evaluating a variety of Electronic Warfare (EW) technologies that can be combined with stealth UAVs which can enter enemy airspace covertly and conduct EW operations at close range. It is likely Russia will develop and produce more stealth UAVs outfitted with EW capability.

Discussion:

Russia has been testing and evaluating a variety of EW technologies including those that operate at close range and those that can potentially function at hundreds of kilometers. [H] Due to operations in Georgia, Ukraine and Syria, Russia has tested close to 200 different military technologies, several EW platforms and capabilities such as the Leer-3 EW platform — a UAV-carried, cell tower-suppressor delivered via a team of two to three Orlan UAVs. [H] The Russian military tested, in Syria especially, the Krasuha and Moskva EW systems, along with the Leer-3. [H]

The Russian unmanned Sukhoi S70 sixth generation fighter is designed as a supersonic stealth plane, with a lot of hardware and hard points, see Figure 43. [M] It is highly likely in 2030-2040 Russian will be able to operate the S-70 Okhotnik with EW functions and units, increasing intelligence, surveillance, and reconnaissance roles. The goal for manned and unmanned vehicles is to be able to carry a mix of sensors and weapons, as required for the mission, and to be able to rapidly integrate new systems and munitions. [H] Russia's growing technological advances in EW will allow its forces to jam, disrupt and interfere

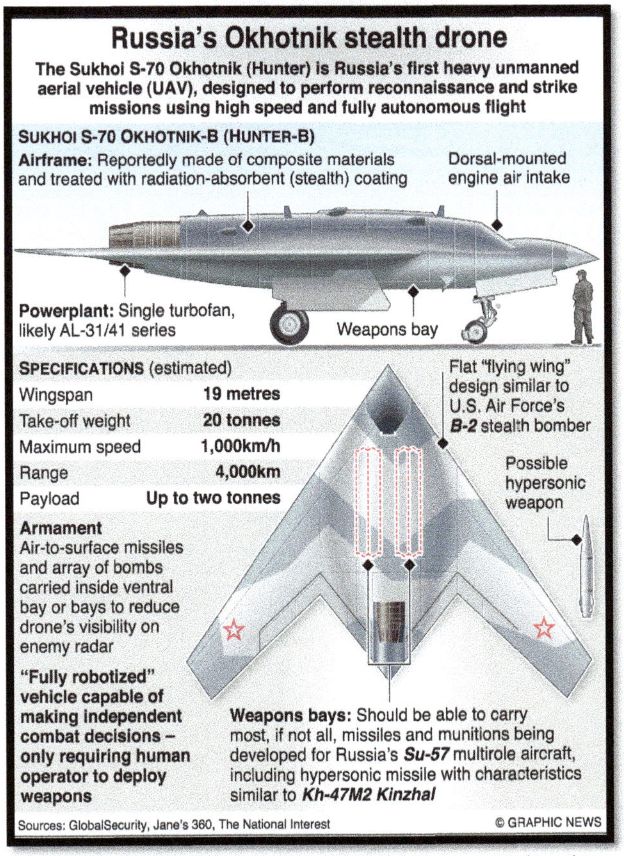

Figure 43: Russia Stealth Drone, the S-70 Okhotnik. As seen above in the photo reportedly Russia's Okhotnik-B stealth attack drone picture leaked. The photo depict a tractor towing the apparently roughly 50-foot-wide unmanned aerial vehicle along a snow-ringed runway at an airfield in Novosibirsk in southern Russia. Source: Janes, Task and Purpose 2019. Can View at: https://taskandpurpose.com/military-tech/photos-of-russias-new-stealth-drone-just-leaked-heres-what-you-should-know

with its adversaries' communications, radar and other sensor systems, Unmanned Aerial Vehicles (UAVs) and other assets, thus negating advantages conferred by the technological edge of the U.S. or its allies. [M] Russia is preparing to confront its adversaries, be it in the air, maritime, land or cyber domains. Russia is highly likely developing and deploying an increasingly capable vast array of EW systems as "force enablers and multipliers". [H] Russia are developing the foundation for what they call a "layered defense," where and under what circumstances EW systems are paired with the early warning radars and air defense systems like Pantsir-S.. [M]

As shown in Figure 44, the EW Forces in Russian operations has increased their support and involvement in operations. Originally designed, EW forces in Russia were a

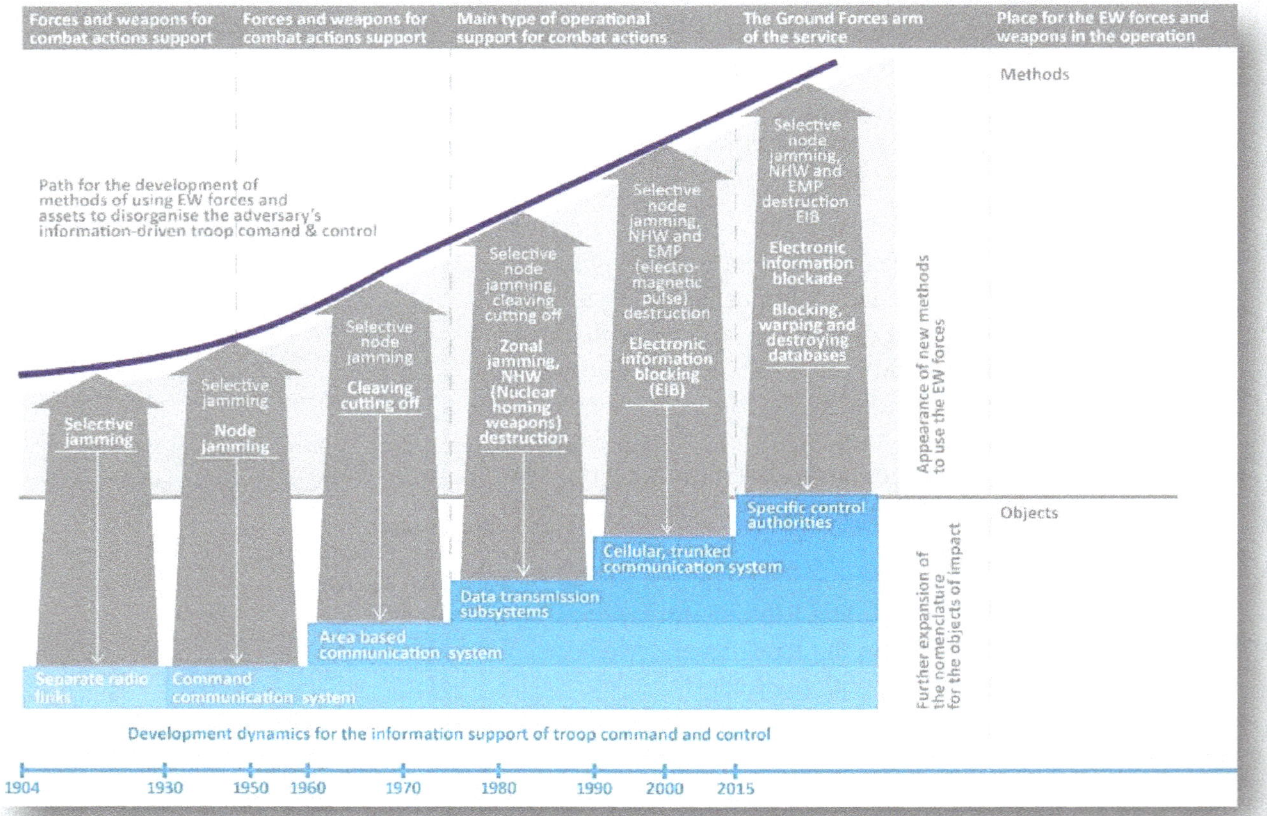

Figure 44: Illustrates since 2015 Russian EW Forces role in Russian operations. The extent to which EW as playing more than a supporting role; if correct, and with greater state funding, by 2025 or later, the EW Forces could emerge as a new combat arm with a pivotal role in military operations. Source: International Center for Security and Defense, Estonia. Can be viewed at: https://icds.ee/wp-content/uploads/2018/ICDS_Report_Russias_Electronic_Warfare_to_2025.pdf.

supporting role, recently and likely in the future Russia EW forces will take more of a leading combat role.[M] Russia's interest in and use of EW is part of a wider effort by Moscow to adopt and strengthen its network-centric capability, which focuses upon C4ISR integration. Russia is already fielding automated command and control (C2) systems that are feeding into EW capability. For example, the Baikal-1ME brigade/regiment-level automated system is interoperable with systems used by EW units. [H] Moreover, these are highly mobile, rendering them difficult to locate. Such developments allow, for instance, Russian forces to establish a highly integrated air defense network and thus improve response times, promote situational awareness and enhance coordination between force elements. [M] Roger McDermott a senior research fellow in Eurasian Military Studies at King's College in London indicates that Russian EW is found throughout every arm and branch of service, making it nearly impossible to avoid. Moreover, all of Russia's combat arms are well-honed from years of electronic combat experience. [M] According to James Faist, the Pentagon's Under Secretary for Defense Research and Engineering, Russia's most powerful EW capabilities reside within its land forces. [M]

According to one of the leading electronic surveillance companies, Era, it is highly likely that sensor and system technologies that collects radio frequency sources across a significant bandwidth range with multiple remote stations and a central processing station, can detect, and pinpoint, signals to 400 kilometers and possibly beyond will be required to analyze and target future EW threat signatures. These signals can be anything from radars, data links, as well as identification friend or foe transponders, meaning even stealth aircraft are visible. [H]

According to Russian Ministry of Defense press releases, EW units regularly conduct training together with other types of unit, such as Signals Troops units and Air Defense radio technical units. Furthermore, according to Jonas Kjellen at the Swedish Defense Research Agency, it is also becoming more common for Russian EW units to train and deploy together with chemical, biological, radiological and nuclear units, and engineers units to be used in protection and concealment against air strikes. For example, in a July 2017 exercise, EW and engineering methods of jointly providing concealment and camouflage were worked out and used as measures against massive attacks from the air. [H]

In 2019 a multifunctional aviation-based electronic warfare/suppression complex was developed as part of the Khibiny R&D project by the Kaluga Radio Engineering Research Institute. [M] Despite these advances, many of the Russian advances in EW have simply involved finally introducing systems that were in the R&D stage since the late Soviet period. [M] It is highly likely that in 2030 Russia's manned and unmanned platforms will have EW equipment for radio reconnaissance and radio suppression, capable of intercepting and suppressing a wide range of radio signals, from cell phones to aircraft and ground-based radars and EW systems. [M]

Analytic Confidence:

The analytic confidence in this estimate is *moderate*. The reliability of the sources available on this topic were above average with several high-quality sources available for the estimate. The sources available did tend to corroborate each other and analyst collaboration was very strong. The estimates contained within the sources are good but given the length of time frame of the estimate, this report is sensitive to changes in market and economic conditions.

Authored by: Sam Smith

Russia Likely to **Slowly Increase Its** Robust Hypersonic Weapons Program By 2030 That Can Evade Current U.S. Air Defenses

Executive Summary:

Due to swift developments across multiple hypersonic technologies, Russia is likely to slowly increase its hypersonic capabilities in the next 10 years. In December 2019, Russia's defense ministry announced it has tested and put nuclear-armed hypersonic weapons into combat duty, allowing Russian President Putin to claim that Russia is the first country armed with hypersonic weapons. Despite its military spending being a fraction of the U.S., Russia will likely develop a robust, hypersonic weapon arsenal.

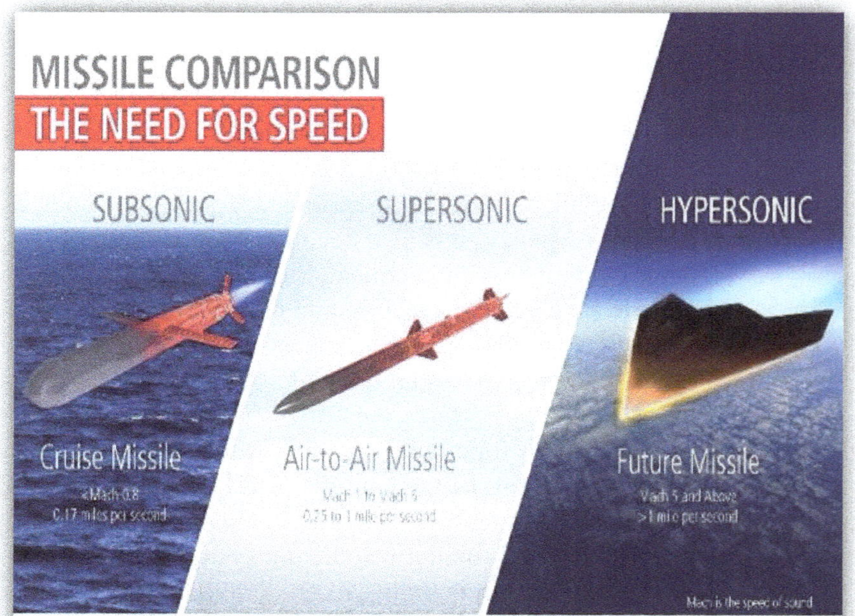

Figure 45: Example of Hypersonic weapons, the future battlespace [H]

Raytheon

Discussion:

Russia is actively experimenting with hypersonic technologies and testing several hypersonic weapons to likely increase its hypersonic program. As illustrated in Figure 45, examples of future hypersonic weapons are likely to advance in size, shape and speed. Russia is attempting to develop hypersonic weapons that feature improved guidance and anti-jam protection that will likely challenge U.S., NATO air defense systems. [H]

Due to swift developments across multiple hypersonic technologies, Russia is likely to slowly increase its hypersonic capabilities in the next 10 years. According to the Russian Defense Minister, in 2019 Russia developed a new intercontinental weapon that can fly 27 times the speed of sound, he further described the Avangard hypersonic glide vehicle as a technological breakthrough that is launched atop an intercontinental ballistic missile, but unlike a regular missile warhead that follows a predictable path after separation it can make sharp maneuvers in the atmosphere en route to target, making it much harder to detect and intercept. [M] Rapidly growing capabilities of hypersonic munitions technology enable global strike capabilities much faster than conventional means. See Figure 46 for Russia's hypersonic weapons.

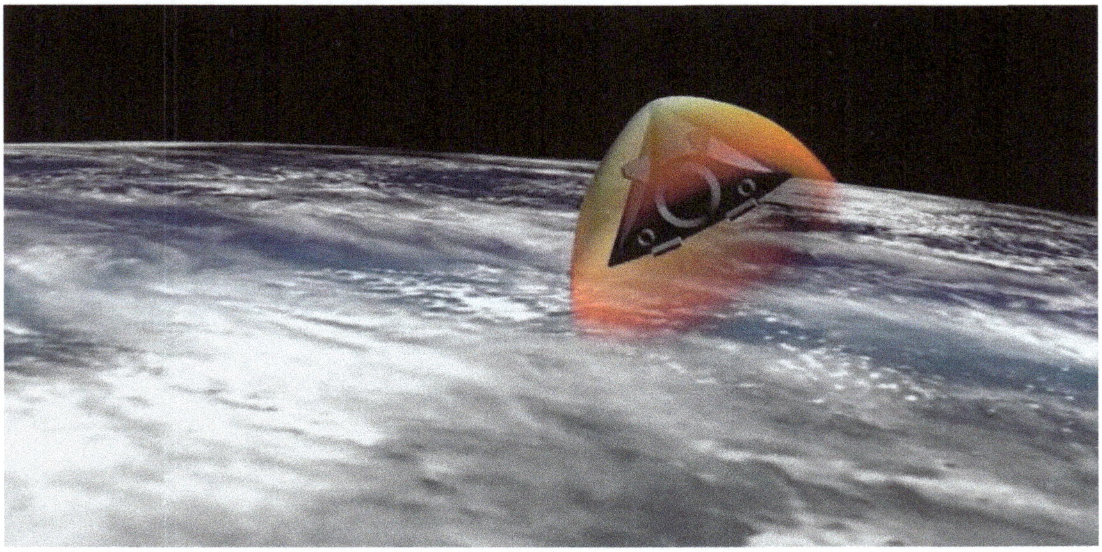

Figure 46: Avangard (Hypersonic Glide Vehicle). Source: U.S. Missile Defense Advocacy Alliance.[M]

Russian/US Designation	Avangard/Objekt 4202, Yu-71, and Yu-7
Role and Mobility	Hypersonic Glide Vehicle; Carried on ICI
Designer/Production	Russian Strategic Missile Forces
Range	Unknown
Warhead Type and Weight	Hypersonic, Nuclear, Conventional/Unkr
MIRV and Yield	Unknown
Guidance System/Accuracy	Unknown
Stages/Propellant	N/A
IOC/Retirement	2018/2019; N/A

Russian President Vladimir Putin noted that the weapon's ability to change both its course and altitude en-route to a target makes it immune to interception by the enemy. [M] Only Russia has nuclear hypersonic capability on an intercontinental-range missile, the Avangard, which Russia declared operational at the Dombarovsky launch site in December 2019. [H] Russia's new hypersonic weapons program includes a guidance system that is being designed to work in tandem with other missiles. See Figure 47.

"It's a weapon of the future, capable of penetrating both existing and prospective missile defense systems," the Russian president stated in December 2019. [M] Due to the frequency and visibility of Russian and Chinese testing, there is a growing sense that the United States is losing the arms race in hypersonic weapons development . [H] In 2018 the Director of the Defense Intelligence Agency stated: "Developments in hypersonic

Figure 47: *Bullet-shaped interceptors defend the United States against attacking hypersonic weapons in an artist's concept. Such defenses remain hypothetical as illustrated by DARPA.* [M]

propulsion will revolutionize warfare by providing the ability to strike targets more quickly, at greater distances, and with greater firepower." [H]

Russia is likely testing hypersonic weapons that can defeat air defense systems as illustrated in Figure 48. Russia recently unveiled a weapon called the Kinzhal, and the Zircon cruise missile which reportedly travels between 3,800mph and 4,600mph – five to six times the speed of sound, and puts Russia several years ahead of the United States. These missiles are reported to reach

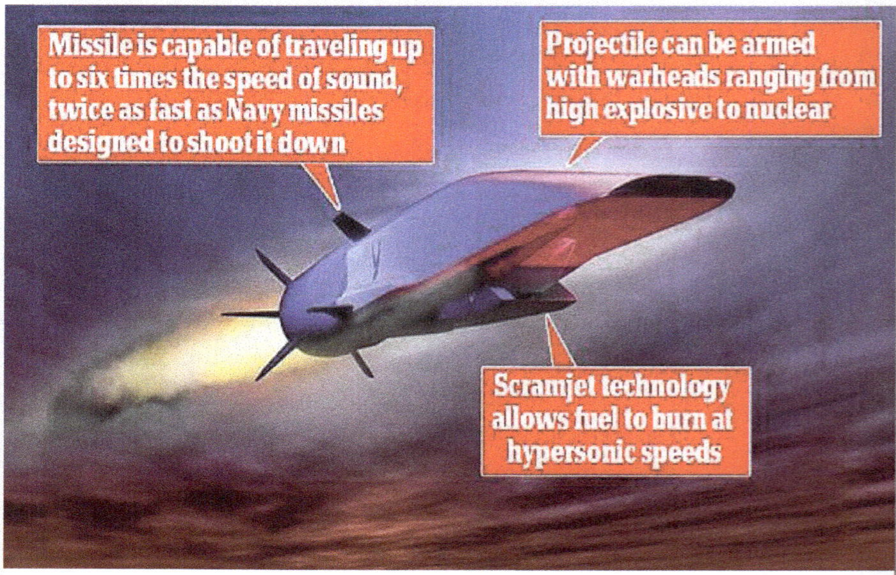

Figure 48: Picture above is an image reportedly of a Russian Zircon cruise missile. Russia successfully tests 'unstoppable' 4,600 mph hypersonic weapon that is faster than any global anti-missile system. *Source:* https://worldofweapon.wordpress.com/2017/06/01/russia-successfully-tests-unstoppable-4600mph-hypersonic-weapon-that-is-faster-than-any-global-anti-missile-system/ [L]

Mach 10 under its own power, and another that is boosted by a rocket to an astonishing Mach 27. [L] Due to this advancement, Russia would enable the Russian jets to attack enemy targets from distances of more than 2,000 km, protecting them from air defense systems. [M]

According to Army technology reports Russia has the ninth largest country defense budget and has allocated RUB 2.9t ($46.4bn) to its defense budget, which excludes allocations to the National Guard, the Border Guard Service, and other confidential line items. According to the Russian Ministry of Defense, plans to upgrade its strategic nuclear assets, submarines and surface vessels, aircraft and helicopters, as well as its aerial capabilities through the induction of new fighters. The modernization will likely support Russia's major acquisition programs, to purchase fighter jets and to further develop hypersonic weapons and missiles that could be used on MiG-31 jet fighters. [H] Russia have reportedly partnered with India, splitting the development costs would allow both countries to add to their capabilities. [M] Despite Russia's planned advancement in its hypersonic program, most of the research and development necessary for these weapons has probably been completed and they are unlikely to be acquired in large quantities so will not involve sizeable additional expenditure. [M]

Analytic Confidence:

Analytical confidence for this estimate is *moderate*. Sources varied on timeframe and capability, despite this, sources did corroborate Russia's intent and resourcing to achieve this development.

Authored by: Sam Smith

IRAN

Iran Estimative Key Findings

Authored By: Rafael Duran

Based on Iran's modernization plan, defense spending, regional strategy, and research analysis that evaluated 19 capabilities against 22 possible signatures, it is highly likely that five signatures— doctrinal/non-technical, physical, digital, and electronic —will both impact the most capabilities and become harder, on average, to detect in the 2030-2040 time frame (See figure 49 below).

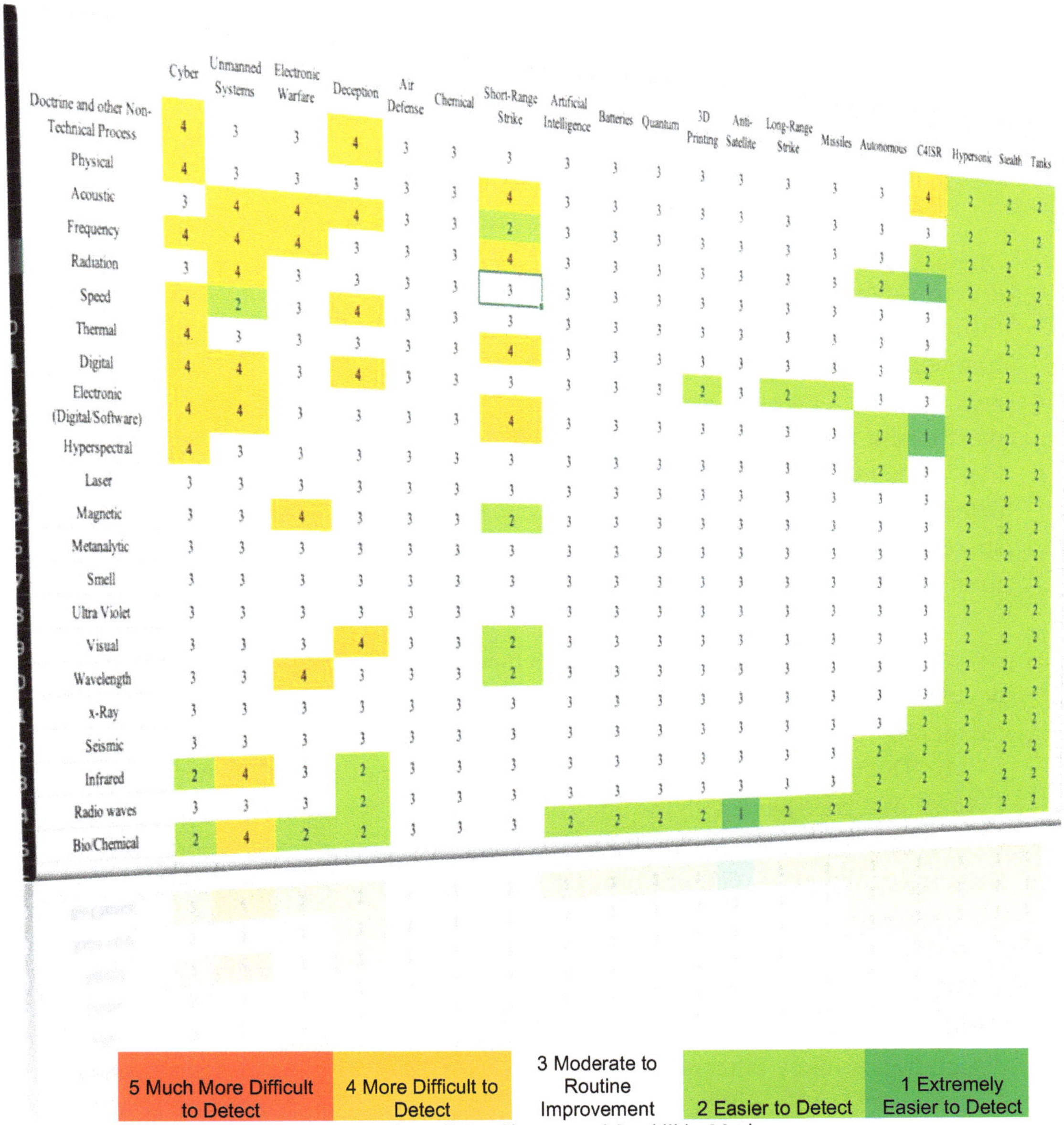

	Cyber	Unmanned Systems	Electronic Warfare	Deception	Air Defense	Chemical	Short-Range Strike	Artificial Intelligence	Batteries	Quantum	3D Printing	Anti-Satellite	Long-Range Strike	Missiles	Autonomous	C4ISR	Hypersonic	Stealth	Tanks
Doctrine and other Non-Technical Process	4	3	3	4	3	3	3	3	3	3	3	3	3	3	3	4	2	2	2
Physical	4	3	3	3	3	3	4	3	3	3	3	3	3	3	3	3	2	2	2
Acoustic	3	4	4	4	3	3	2	3	3	3	3	3	3	3	3	2	2	2	2
Frequency	4	4	4	3	3	3	4	3	3	3	3	3	3	3	2	1	2	2	2
Radiation	3	4	3	3	3	3	3	3	3	3	3	3	3	3	3	3	2	2	2
Speed	4	2	3	4	3	3	3	3	3	3	3	3	3	3	3	3	2	2	2
Thermal	4.	3	3	3	3	3	4	3	3	3	3	3	3	3	3	2	2	2	2
Digital	4	4	3	4	3	3	3	3	3	3	2	3	2	2	3	3	2	2	2
Electronic (Digital/Software)	4	4	3	3	3	3	4	3	3	3	3	3	3	3	2	1	2	2	2
Hyperspectral	4	3	3	3	3	3	3	3	3	3	3	3	3	3	2	3	2	2	2
Laser	3	3	3	3	3	3	3	3	3	3	3	3	3	3	3	3	2	2	2
Magnetic	3	3	4	3	3	3	2	3	3	3	3	3	3	3	3	3	2	2	2
Metanalytic	3	3	3	3	3	3	3	3	3	3	3	3	3	3	3	3	2	2	2
Smell	3	3	3	3	3	3	3	3	3	3	3	3	3	3	3	3	2	2	2
Ultra Violet	3	3	3	3	3	3	3	3	3	3	3	3	3	3	3	3	2	2	2
Visual	3	3	3	4	3	3	2	3	3	3	3	3	3	3	3	3	2	2	2
Wavelength	3	3	4	3	3	3	2	3	3	3	3	3	3	3	3	3	2	2	2
x-Ray	3	3	3	3	3	3	3	3	3	3	3	3	3	3	3	3	2	2	2
Seismic	3	3	3	3	3	3	3	3	3	3	3	3	3	3	3	2	2	2	2
Infrared	2	4	3	2	3	3	3	3	3	3	3	3	3	3	2	2	2	2	2
Radio waves	3	3	3	2	3	3	3	3	3	3	3	3	3	3	2	2	2	2	2
Bio/Chemical	2	4	2	2	3	3	3	2	2	2	2	1	2	2	2	2	2	2	2

Figure 49 Iran Signature and Capabilities Matrix

1. The future Iran threat signature most difficult to detect 2030-2040 is likely digital from the cyber and unmanned systems capabilities. In fact, only 7% (29/418) of Iran's threat signatures distributed across 19 capabilities and 22 potential signatures will likely become difficult to detect between 2030 and 2040. 60% (287/418) of future Iran threat signatures distributed across 19 capabilities and 22 signatures will be moderate to routine (incremental changes only) to detect, and 20% (81/418) of the signatures associated with these 19 capabilities are likely to become easier to detect in the same period. Driving these changes are several key facts. Specifically:

   a. In accordance to the 2020 Index of Economic Freedom, Iran's economy is faltering and unpredictable oil prices yield less than expected return on investments likely limiting future defense modernization.
   b. According to the Balance contributor, Kimberly Amadeo, the U.S. decision to withdraw from the Joint Comprehensive Plan of Action (JCPOA) is creating and coercing Iran to take extreme measures focusing on easily acquired technology.
   c. The vast availability of affordable cyber, unmanned systems, and missiles technology, couple with Iran's indigenous resources present Iran with prompt operational and strategic respond systems in light of regional and global pressures

2. Iran's cyber digital signatures are highly likely to improve by twofold becoming difficult to detect and acquired through the acquisition of foreign sources technology. Iran will continue to threaten private and government networks worldwide within the next 10 years. Out of 19 capabilities and 22 signatures, Iran's cyber capabilities will mostly emit the future threat signatures of doctrine, physical digital, electronic, and hyperspectral topping 6% of all future Iran's treat signatures likely becoming difficult to detect. Iran will improve cyber warfare capabilities employing legitimate and illicit strategies to obtain the latest technology including black-market acquisitions, and the foreign employment of students for academic enrichment.

a. (Doctrine/Digital/Electronic/Hyperspectral) Iran is making every effort to obtain, develop, and employ cyber offensive and defensive technology in accordance with a recent report from the Jane's Sentinel Security assessment

b. (Doctrine/Physical) The Jane's report added Iran's establishment of a Supreme Council of Cyberspace with the objective to conduct cyber proliferation

c. (Doctrine/Physical According to Israeli news sources, and reported in Jane's, the Iranian Revolutionary Guard Corps (IRGC) inaugurated the first Cyber Defence Centre headquarters

3. Iran's UAVs mass production is moving at an accelerated pace ahead of all Middle East Arab countries likely ensuring that its future UAVs signature will be difficult to detect. Iran is likely to increase the acoustic, frequency, digital, electronic, and infrared signatures of UAVs within the next 15 years. Driving these changes are several key facts:

a. As per the Jane's Weekly Gulf States publication Iran is increasing UAVs radar cross section and electromagnetic signatures.

b. Jane's and the Associated Press (AP) reports indicate that numerous successful Iranian UAVs involved in the Syria conflict and the recovered drones in Saudi Arabia with Iran's manufacturing marks.

c. AP reports on Iran's ability to employ armed UAVs against insurgent forces in Syria.

4. With mounting regional pressure, Iran currently focus on the immediate development of short-range ballistic missiles (SRBM). However, Iran recognizes the importance of strategic assets to project global power and continues to make significant efforts towards the development of Intercontinental Ballistic Missile (ICBM). Iran's SRBM signatures including acoustic, frequency, thermal, and electronic will likely become more difficult to detect in the next 10 to 15 fifteen years. Key findings supporting the improvements on the SRBM include:

a. A recent survey from the Jane's Sentinel Security Assessment reveals that Middle East regional states find Iran's ICBM propagation and testing as the top strategic concern

b. Iran's future ICBM threat signatures are highly likely to remain constant with those of near peer competitors, however, drastic improvements in SRBM is the norm

a. c. The Center for Strategic National Studies (CSNS) affirmed that Iran's SRBM program will continue to mature and take precedence over global strategic power projection.

5. Figure 50 represents Iran's future threat signatures when the capabilities (lowest, less confident threat capabilities) are removed. 28 out of 308 (2%) potential signatures across 14 capabilities including frequency, electronic, acoustic, doctrine, physical, hyperspectral, hypersonic, speed, and digital signatures scored the most difficult to detect

	Cyber	Unmanned Systems	Electronic Warfare	Deception	Short-Range Strike	Air Defense	Chemical	Artificial Intelligence	Batteries	Quantum	3D Printing	Anti-Satellite	Long-Range Strike	Missiles
Frequency	4	4	4	3	4	3	3	3	3	3	3	3	3	3
Electronic (Digital/Software)	4	4	3	3	4	3	3	3	3	3	3	3	3	3
Acoustic	3	4	4	4	2	3	3	3	3	3	3	3	3	3
Doctrine and other Non-Technical Process	4	3	3	4	3	3	3	3	3	3	3	3	3	3
Physical	4	3	3	3	4	3	3	3	3	3	3	3	3	3
Thermal	4	3	3	3	4	3	3	3	3	3	3	3	3	3
Hyperspectral	4	3	3	3	3	3	3	3	3	3	3	3	3	3
Radiation	3	4	3	3	3	3	3	3	3	3	3	3	3	3
Speed	4	2	3	4	3	3	3	3	3	3	3	3	3	3
Laser	3	3	3	3	3	3	3	3	3	3	3	3	3	3
Magnetic	3	3	4	3	2	3	3	3	3	3	3	3	3	3
Metanalytic	3	3	3	3	3	3	3	3	3	3	3	3	3	3
Seismic	3	3	3	3	3	3	3	3	3	3	3	3	3	3
Smell	3	3	3	3	3	3	3	3	3	3	3	3	3	3
Ultra Violet	3	3	3	3	3	3	3	3	3	3	3	3	3	3
Visual	3	3	3	4	2	3	3	3	3	3	3	3	3	3
Wavelength	3	3	4	3	2	3	3	3	3	3	3	3	3	3
x-Ray	3	3	3	3	3	3	3	3	3	3	3	3	3	3
Digital	4	4	3	4	3	3	3	3	3	3	2	3	2	2
Infrared	2	4	3	2	3	3	3	3	3	3	3	3	3	3
Radio waves	3	3	3	2	3	3	3	3	3	3	3	3	3	3
Bio/Chemical	2	4	2	2	3	3	3	2	2	2	2	1	2	[illegible]

5 Much More Difficult to Detect	4 More Difficult to Detect	3 Moderate to Routine Improvement	2 Easier to Detect	1 Extremely Easier to Detect

Figure 50 Iran Signature and Capabilities Matrix w/ Focus on Top Signatures and Projected Capabilities

in the next 10 to 20 years. It is likely that it will be significantly difficult to detect Iran UAVs signatures in 2030-2040 due to the continue production of the Iranian increase in radar cross section, acoustic, surface temperature, and electromagnetic technology prompting U.S. forces to develop and apply multispectral systems to capture the myriad of potential signatures.

7. The U.S. Army will likely need Measurement and Signature Intelligence (MASINT) sensors and systems that can detect, track, identify or describe the distinctive

characteristics of Iran fixed or dynamic visual, thermal, radio waves, speed, digital and non-technical signatures.

8. 100% (22/22) of future Iran stealth fighter potential threat signatures will likely become easy to detect between 2030 and 2040. Iran's first stealth fighter demonstrated a slow progression in air power and stealth defense capabilities, and proof of a highly likely ineffective stealth fighter signature reduction within the next 15 years.

9. Aside from slow technological advantages, the Qaher F313 demonstrated Iran's first attempt of stealth fighter as a means of deterrence. BBC report noted that Israel National Defense experts disregarded the F313 as a competitive stealth fighter adding that Iran's has not developed convincing technology.

Iran Likely to Reduce Cyber Warfare Digital Signatures by Twofold Increasing Cyber Technology Through Foreign Sources within the Next 10 Years

Executive Summary:

Iran cyber warfare digital signatures are highly likely to improve its anti-detection technology by twofold through the acquisition of foreign sources technology and will continue to threaten private and government networks worldwide within the next 10 years. Iran will improve cyber warfare capabilities employing legitimate and illicit strategies to obtain the latest technology including black-market acquisitions, and the foreign employment of students for academic enrichment. Iran's future Cyber warfare capabilities will decrease the U.S.'s ability to detect cyber signatures because the evolving cyber intelligent malware that facilitates the penetration of intrusion detection and prevention systems. Due to major U.S. and Israeli cyber-attacks coupled with threats of domestic rebellion, Iran intensified the research of cyberwarfare technology. Despite Iran's significant improvements in cyber warfare capabilities, Iran will continue to be vulnerable to external cyber-attacks from developed nation states.

Discussion:

As depicted in Figure 51, the Concentric 2019 Report lists Iran at the top overall risk of any country in the Middle East. [H] The risk is defined in terms of capability, intent, and presence. In a scale of 1-5, with 1 measuring as the highest threat, Iran ranks number 1 in the Middle East. When comparing Iran's cyber capabilities to those of leading nations, the Concentric 2019 Track-Report posted that "American software firm Cylance suggests that Iran has reached a cyber capability close to that of China or Russia." [H]

	Overall Risk	Capability	Intent	Presence/Reach
MIDDLE EAST				
Iran	3.500	3.167	3.708	3.625
Israel	2.832	3.433	2.500	2.563
Qatar	2.133	2.400	1.667	2.333
Saudi Arabia	2.329	2.500	1.875	2.611
United Arab Emirates	2.667	3.000	2.500	2.500

Figure 51 Concentric Advisors

The Iranian Islamic Revolutionary Guard Corps (IRGC) activated the first Iranian Cyberwarfare Command (IRGCCWC) promising cyber warfare technology improvements within the next 10 years. First strong evidence that Iran is likely to improve its cyber technology is simply the direct technological assistance Russia provides. Mike Beck, top Darktrace Analyst emphasized on Russia's national interest to assist Iran in improving and developing cyber warfare to access U.S. systems. [H] Second source contributing to Iran's cyber warfare technology improvement is North Korea who gathers, process, and share significant cyber warfare technology and capabilities that enhance Iran's ability to continue to proliferate cyber warfare. Under these foreign supports, couple with the students' academic collection program, Iran will continue to exploit the improvement of cyber warfare capabilities avoiding detection during the conduct of cyber offensive operations, and to protect itself from future U.S. cyber-attacks. It is in Russia's strategic interests to pursuit the proliferation of cyber capabilities in the Middle East utilizing Iran as the center of gravity. Third evidence, today's Iran cyber warfare capabilities are far more advance than eleven years ago in 2009. At that demonstrated rate of improvement, Iran's cyber capabilities will be far more advance in the next ten years including a much lower signature capable of avoiding detection and penetrating fire walls. [H]

Iran's recent attacks demonstrated continuous cyber warfare research and improvements. Extensive Cyber warfare research improved Iran's ability to achieve strategic objectives without employing the military instrument of national power; additionally, experts believe that Iran's cyber-attacks against the Saudi Arabia gas company Aramco served as evidence of unprecedented efforts for continuous future improvement of cyber warfare technology. Iran employed cyber warfare to demonstrate its ability to offensively project power worldwide while minimizing the cyber signature though the employment of sophisticated penetrating malware. Iran actively retaliated with numerous cyber-attacks demonstrating the capabilities for covert and open cyber-attacks against Saudi Arabia and the U.S. with strategically disabling attacks that rendered Aramco ineffective and disabling banks communication networks in the U.S.[H]

Iran upgraded the level of cyber-attacks and future cyber warfare developments will surpass the technology of "recent cyberattacks, including the 2017 compromise of 90 British parliamentarian's email accounts and 2019 phishing attempts targeting the U.S. Department of Energy, believed to originate from operational cells in the IRGC lend credence to Cylance's claim." [H] United Nations (UN) Resolution 2231, adopted by the Security Council at its 7488th meeting, on 20 July 2015 will allow Iran to conduct licit weapons and technology purchases after summer 2020. [H] Freedom to join the world weapons market will allow Iran to continue the improvement of cyber warfare technology. Demonstrating hope to continue with technological improvements and convincingly including cyber warfare research and development, "President Hassan Rouhani said on Monday Iran would regain access to the international arms market… if it stuck to its 2015 nuclear deal with the world powers..." [H] Cyber security scholars speculated that another Iranian motivation to improve its cyber warfare capabilities in the future "follows a succession of cyber-attacks on Iranian networks including the revelation that U.S. President Barack Obama authorized the Stuxnet computer virus, which was developed with an Israeli intelligence unit, to attack Iran's Natanz uranium enrichment facility."[H] The emergence of the Flame malware virus coerced Iran's to increase cyber warfare capabilities as a form of retaliation since "Iran's oil infrastructure also experienced a series of cyber-attacks in April, was targeted by the "Flame' virus in May, and, this month, security experts at Kaspersky Lab have reportedly uncovered an ongoing cyber-espionage campaign against the country which features malware written partly in Persian."[22]

Although Iran has made remarkable developments in cyberwarfare, others argue that Iran's will remain behind U.S. in cyber capabilities. According to a 2019 Center for Strategic and International Studies report, "Iran has rapidly improved its cyber capabilities, it is still not in the top rank of cyber powers, but it is ahead of most nations in strategy and organization for cyber warfare."[H] Iran continues to improve its cyber warfare capabilities and "Iran has a good appreciation for the utility of cyber as an instrument of national power. . .its extensive experience in covert activities helps guide its strategy and operations using cyber as a tool for coercion and force, and it has created a sophisticated organizational structure to manage cyber conflict."[H] Another probable purpose of Iran's future aggressive development of cyber warfare capability is that "what Iran's leaders fear most, however, is their own population and the risk that the internet will unleash something like the Arab Spring."[H]

Analytic Confidence:

Analytical confidence is *high*. Adequate time provided for the availability and research of ample credible sources. All sources mutually corroborated with the analysis. Due to the factor of time, the rapid pace of technology may potentially alter forecasting thereby shortening the range of development.

Authored by: Rafael Duran

---

[22] "Iran to establish new cyber defence headquarters", UK, in Jane's Defence Weekly 20 Jul 12 (accessed through Jane's Group UK Limited): "The announcement follows a succession of cyber-attacks on Iranian networks including the revelation that U.S. President Barack Obama authorized the Stuxnet computer virus, which was developed with an Israeli intelligence unit, to attack Iran's Natanz uranium enrichment facility. Iran's oil infrastructure also experienced a series of cyber attacks in April, was targeted by the 'Flame' virus in May and, this month, security experts at Kaspersky Lab have reportedly uncovered an ongoing cyber-espionage campaign against the country which features malware written partly in Persia."

Iran's Radar, Acoustic, Temperature, and Electromagnetic Spectrum Signatures of Unmanned Arial Vehicles Will Likely Become Difficult to Detect Within the Next 15 Years

Executive Summary:

Iran's Unmanned Arial Vehicles (UAV)s mass production is moving at an accelerated pace ahead of all Middle East Arab countries likely increasing its UAVs signature. Iran is likely to increase the radar, acoustic, temperature, and electromagnetic signatures of UAVs within the next 15 years. Due to inadequate air power and pressure from adversaries' superior air forces, Iran develops UAVs as an affordable and advantageous strategic instrument. Despite international embargos, Iran has demonstrated the ability to sustain the substantial production of UAVs. However, Iran is not the lead in UAV technology in the Middle East. Israel's UAVs technology is at forefront of UAVs research and development (R&D). Iran's future increase in the employment of UAVs will require the U.S. and allies to generate technology to identify the myriad of UAVs signatures in the battlefield. Current technology is limited in identifying UAVs signatures which could allow Iran to disrupt, elude, or completely evade military defense systems.

Discussion:

Iran has effectively employed UAVs regionally in support of Syria, Yemen, and non-state organizations such as Hamas and Hezbollah. Iran has also effectively employed UAVs to attack targets of interest within Saudi Arabia. The ability to detect UAVs is dependent on

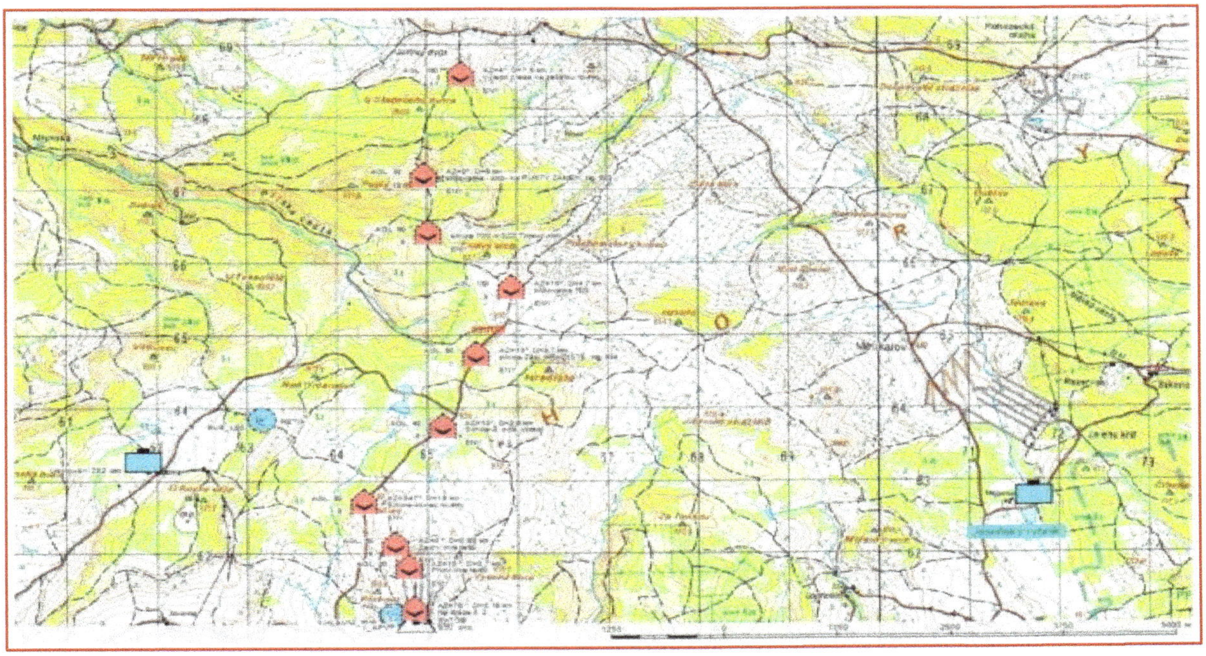

Figure 52 UAVs Signatures ID

the type of signature emitted and the composition of the UAV. Figure 52, from the Department of Air Defense (DAD), Faculty of Military Technology (FMT), University of Defense, Kounicova, Czech Republic, Jan Farlik depicts a study conducted on multispectral detection of commercial unmanned aerial vehicles, published on April 2019, [H] represents a combination of different types of observers, blue icons, including naked eye, telescope, and electronic devices. The study observed that employing a combination of systems increase the effectiveness of detection because UAVs project a multitude of signatures including radar cross section, noise level, surface temperature, optical, acoustic, and electromagnetic spectrum bands that will require advance sensor detection systems as the UAVs technology improves. Figure 53 [H], Department of Air Defence, Faculty of Military Technology, University of Defence, Czech Republic, clearly demonstrates a range of multispectral detection of commercial unmanned aerial vehicles. The lowest range of detection for an UAV commercial signature is the unarmed eyes reaching up to 250 meters, and the highest range of detection with combined radar sensors is up to 5,000 meters range. [H]

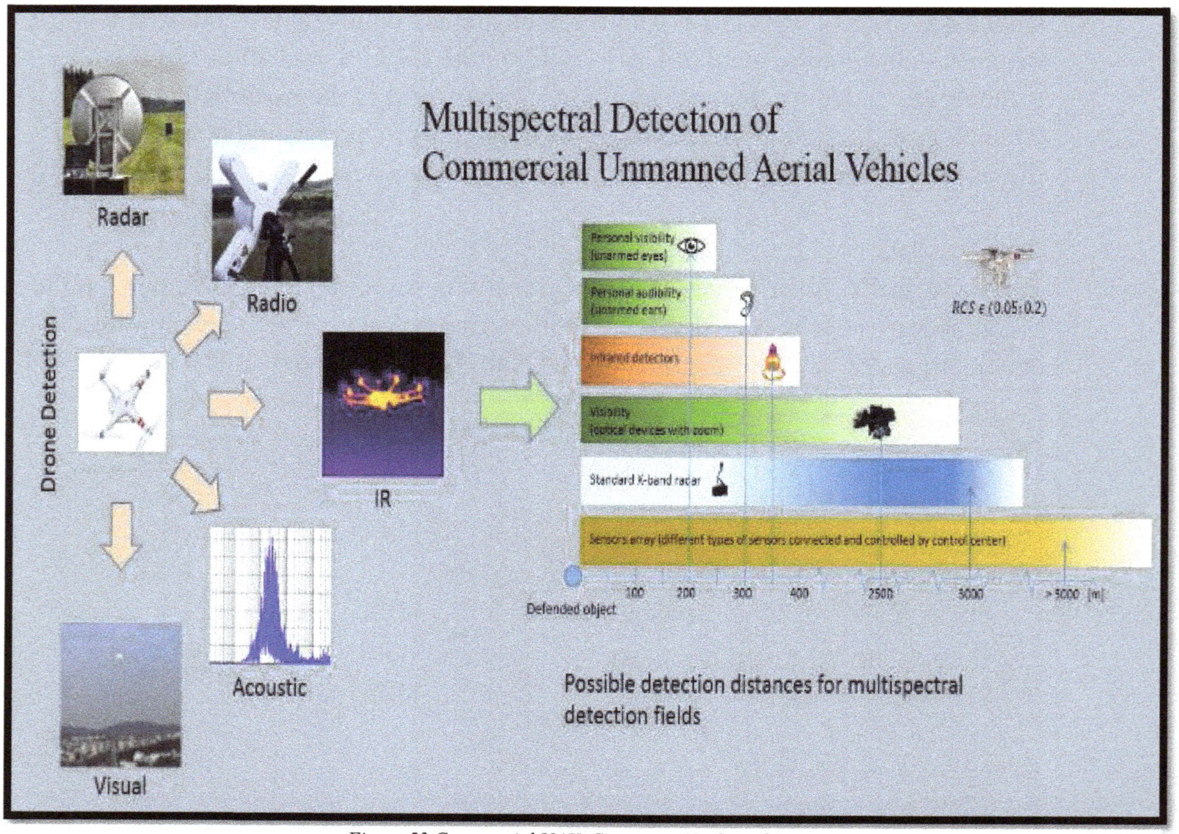

Figure 53 Commercial UAVs Signatures Evaluated Range

Strong indicators of Iran's increase of UAVs radar cross section and electromagnetic signatures are the numerous successful Iranian UAVs involved in the Syria conflict. Iran has been capable of arming its UAVs against insurgent forces in Syria. [H] Additionally, Iranian UAVs effectively supported the Houthi rebels, the terrorist organization

Hezbollah, and the downing of an American UAV demonstrating Iran's ability to obtain the technology to build reliable unmanned aerial vehicles. [H] As a future strategic program, "the Iranian state has invested heavily in a number of UAV technologies since the mid-2000s, viewing these platforms as a means to augment the limited capabilities of its air force. Iran has domestically produced UAVs for a number of years and has made advancements in range, payload, and capability commensurate with focused research and development efforts to expand the unmanned fleet." [23] Iran's UAV R&D focused on "the recent airstrike in Syria attributed to Israel [and] has brought to the forefront Iran's intentions of establishing a network of drones (unmanned aerial vehicles) in that country. The project could expand the Islamic Republic's capabilities of gathering intelligence and prepare the groundwork for possible attacks." [M]

However, there are communication restrictions and "the main difficulty facing the Islamic Republic is its lack of a military satellite. Most Iranian aerial vehicles cannot transmit information to their handlers in real time. Any intelligence can be used only after the drones return to base. Some of them can communicate through radar but this applies only to short ranges." [M] Furthermore, the continue production of Iranian UAVs will increase the radar cross section, acoustic, surface temperature, and electromagnetic signatures prompting U.S. forces to develop and apply multispectral systems to capture the myriad of potential signatures. The future development of U.S. anti-drone systems must account for the different types of signatures to balance the economic cost of the drone employed and the detection and defensive system. For instance, a Heritage Foundation study on Iran's growing UAV threat, and a Defense News report alluded that shooting down a cheap armed drone with an expensive missile does not benefit the U.S. economically. [H]

United Nation embargo resolutions appear not to have made an imposing effect diminishing Iran's UAV technological development. U.S. intelligence agencies concur that despite the embargos, Iran will continue to improve its UAVs technology through illicit means across black markets and internal production. Analysts discard Iran's ability to reach world dominant production levels, but "as for the future, Iran's military is likely to maintain its focus on expanding and improving its unmanned fleet. To that end, Cooper says that his Iranian sources tell him that Iran's Defense Industries Organization plans to roll out a UAV-related 'surprise' in September." [H]

---

[23] "Iran – Air Force", London, in Jane's World Air Force 04 Mar 20 (accessed through Jane's Defence Weekly): "The Iranian state has invested heavily in a number of UAV technologies since the mid-2000s, viewing these platforms as a means to augment the limited capabilities of its air force. UAVs have been produced domestically in Iran for a number of years and have advanced in range, payload, and capability that commensurate with focused research and development efforts to expand the unmanned fleet. The Shahed-129 was originally unveiled in 2012 and was reported to have a range of 2,000 km and total endurance of up to 24 hours. In July 2014 the Shahed-129 was first observed operating in Syria, being acknowledged in 2016 by the IRGC as one of the principal assets deployed abroad to support operations in Syria and Iraq. The Shahed-129 has been modified to carry munitions, including the Sadid-345 guided bomb."

Analytic Confidence:

Analytical confidence for this estimate is *high*. Credible and collaborating sources demonstrated Iran's ability to increase its UAVs effectiveness to achieve strategic advantages. Under the current operational environment, sources demonstrated a high tendency for Iran to rapidly advance UAVs research and development.

Authored by: Rafael Duran

Iran Intercontinental Ballistic Missiles Heat Signature Likely to Remain Constant within Next 10 Years

Executive Summary:

With the advanced long-term development of ballistic missiles (BM) and foreign technological support make it likely that Iran will test a successful Intercontinental Ballistic Missile (ICBM) within the next 10 years. Iran's ICBM signatures detection are likely to remain moderate to routine within the next 10 years because it will likely resemble the same signatures as the old Soviet and North Korea Taepo Dong missile launching platforms and rocket capabilities. Due to an antiquated air force, Iran focuses on the need to develop a robust ballistic missile program to deter regional adversaries such as Israel, Western Europeans, and the presence of U.S. military in the Middle East. Despite Iran's prominent ICBM incremental advances, experts concur that Iran's priority is on the development Short Range Ballistic Missiles (SRBM) because of the shorter range require to meet the high demands of imminent regional threats.

Discussion:

In early 1988, during the Iraq/Iran War, China began to provide SRBM technology to Iran influencing a successful BM program. By 2000, the U.S. Central Intelligence Agency (CIA)[11] affirmed that Iran

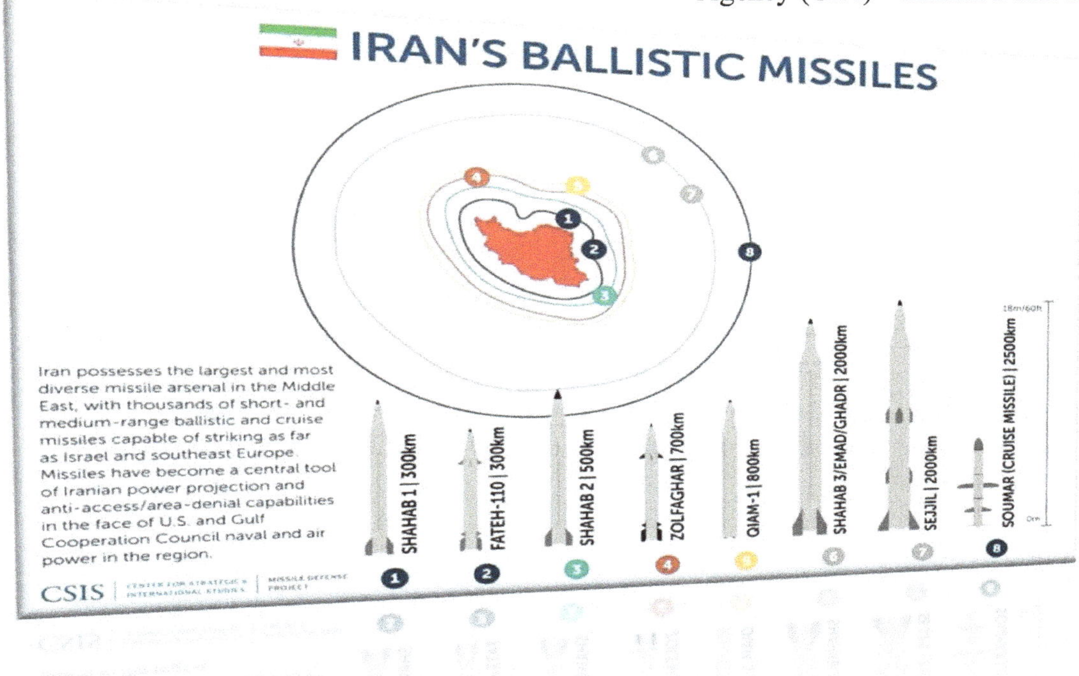

Figure *54* The Shahab SRBM is capable of Western Europe range. Zachary Keck. The National Interest, July 11, 2019

had achieved considerable technological and material progress towards BM production capabilities. According to Janes:

> *"Today, Iran possesses the capability to deploy ballistic missiles (and long-range artillery rockets) against its regional neighbors, as well as Israeli and U.S. forces deployed in the region. It has bolstered this capability with assistance from China, North Korea, and Russia since the 1980s. However, Iran has also sought to reduce its reliance on foreign support and is believed to have developed considerable in-country development capabilities."* [24]

Per the Center for Strategic and International Studies (CSIS), Iran's SRBM program takes priority over the ICBM development due to the imminent regional threats and "seen from an Iranian perspective, Iran is responding to proven threats from its neighbors and the U.S., and its inability to properly modernize its military forces since 1980." [H]

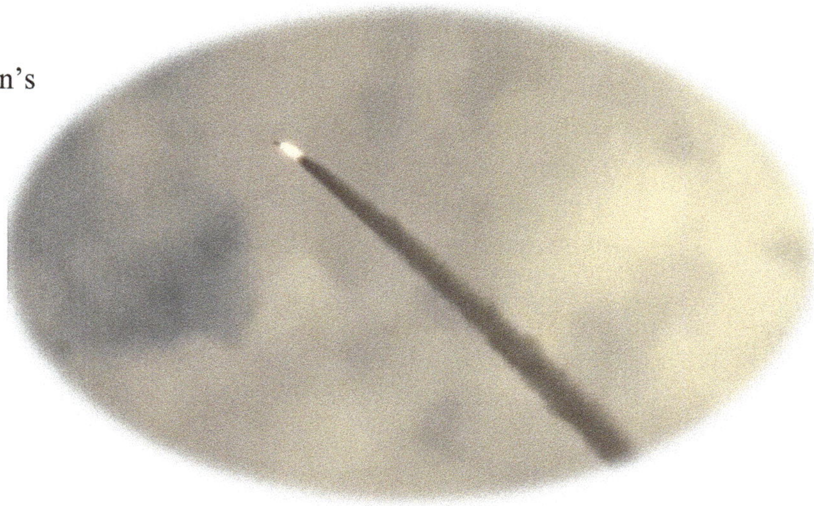

Figure *55* The Shahab SRBM is example
Click Picture to be connected to video (must have internet connection)

Iran continues to engage in the development of ICBM to gain strategic leverage and to increase its global nuclear capability. Although Iran has developed its own SRBM, most of the current and future missile signatures will assimilate the same missile technology from Iran's supporting countries such as China, North Korea, and Russia. North Korea and China illegally deliver BM material necessary for Iran to expand capabilities from short and medium range ballistic missiles to the development of intercontinental ballistic missiles (ICBM). [M] Besides North Korea and China, Russia has been an integral contributor to Iran's SRBM research and development (R&D) enhancing range capabilities and precision fires. [M]

It is likely that if Iran conducts a future ICBM launch, the heat signature will remain constant because "Iran has a well-developed technological and industrial capability to

---

[24] "Iran – Strategic Weapon Systems", UK, in Jane's Group UK Limited 20 Feb 20 (accessed through Jane's Sentinel Security Assessment – The Gulf States, Section Strategic Weapon System): "Iran possesses the most significant ballistic missile inventories in the Middle East, having acquired complete missile systems in the past and subsequently developing infrastructure to build a variety of ballistic missiles indigenously. With military sanctions occurring in the wake of the 1979 Islamic Revolution, Iran's ability to support its Western-supplied combat aircraft waned alongside its ability to conduct long-range strikes. As a means to counter this shortcoming, Iran sought a ballistic missile capability to offer a means of striking adversaries beyond Iran's borders.

build short-range and medium-range missiles on a large scale, but it must still cross a number of technological thresholds concerning stage separation, propulsion systems, re-entry vehicles, and guidance systems before it could successfully test an ICBM. [H] The latest Iranian successful efforts to launch space satellites demonstrated some of the capabilities needed to build ICBMs. [H] Iran's space program increases the ability for the future employment of ICBMs although additional research and development would still be needed to include atmosphere reentry technology. [M] If successful, Iran's placement of additional satellites in low orbit increases strategic defensive and offensive capabilities, and the ability to interrupt satellite communications, but the satellite launching platform signatures will mirror the ICBM launching.

Analytic Confidence:

Analytical confidence for this estimate is *high*. Available adequate time supported the analytical process. There is an abundancy of reliable sources that mostly corroborated with each other.

Authored by: Rafael Duran

Iran Unlikely to Reduce Stealth Fighter Signatures Within the Next 15 Years

Executive Summary:

Iran's first stealth fighter demonstrates a slow progression in air power and stealth defense capabilities, and proof of a highly likely ineffective stealth fighter signature reduction within the next 15 years. Aside from slow technological advantages, the Qaher F313 demonstrated Iran's first attempt of stealth fighter as a means of deterrence. Due to the continuous air power threats from Iran's adversaries, Iran is focused on developing stealth technology; however, continuous U.S. sanction have likely slowed progress. Despite the increased focus to produce stealth fighters, Iran has been stagnant in the prototype phase for 7 years on the first stealth fighter. Experts agreed on the inability of the Qaher F313 design to reduce its signature due to poor avionics designs and unreliable technology.

Discussion:

Critics are divided on the effectiveness of the Qaher F313 assessing Iran's technology as incompetent and infective. Unreliable sources from Iran itself, view Iran's stealth fighter technology as a significant competitive mean that increases the country's defensive and offensive capabilities. [M] According to BBC, during a military weapons expo "President Mahmoud Ahmedinejad said it had [almost all the positive features] of the world's most sophisticated jets. [L] He said the "development of the Iranian nation's military power is... for deterrence and defensive purposes." [L] Independent sources confirm that Iran will continue with the research and development (R&D) of stealth technology. Production of efficient F313 is questionable and research indicate a high signature emission for an aircraft that is intended to evade radars. Iran is making every effort to increase deterrence strategies through the adaptation of existing stealth fighter technology, but strong embargo and monitoring have reduced the ability for fast technological developments and implementations.

Figure 56 Iran Qaher-313 jet prototype (photo credit: Tasnim News Agency)

Iran is strongly pursuing stealth technology to reduce heat signature making every legal and illegal effort to acquire material, because of the current sanctions, and the lack of internal resources to develop competitive stealth fighters. Iranian officials claimed the reverse engineering of a captured advanced U.S. stealth RQ-170 stealth unmanned aerial vehicle likely decreasing heat signature and improving its national strategy. Jane's Defence Weekly reported "the announcement that an Iranian version of the classified UAV has flown was made by the commander of the Islamic Revolution Guards Corps (IRGC) aerospace division, Brigadier General Amir Ali Hajizade, according to the *Tasnim News Agency.*" [25]

Israel National Defense experts disregarded the F313 as a competitive stealth fighter adding "Iran has been under tight sanctions since its war with Iraq and has not been allowed to obtain new hardware or spare parts for its old fleet. Buying Russian or Chinese jet fighters is not a ready option for Iran due to new sanctions imposed over its controversial nuclear programme." [M] Other sources discussed the F313 stealth fighter summarizing ten top capabilities to include "use of radar-absorbent materials in the body, to absorb wave energy and reduce reflection, for a greater stealth effect of Qaher F-313 fighter, and the hot exhaust gas mixes with cold air through the inlet ducts, and gets cooler before it gets out of the exhaust system, to reduce heat effects on the surface of the aircraft." [H] However, there is little to no trust on Iran's ability to reverse engineer the advance stealth technology and reduces thermal signature. Jane's Weekly Defence added that "in addition to this stealthy payload, the Sentinel's airframe and surfaces will also be largely made up of low observable materials and treatments, none of which Iran is believed to be able to manufacture." [26]

Analytic Confidence:

Analytical confidence is high supporting that Iran will likely not develop competitive stealth technology to reduce its thermal signature in the next 15 years. Multiple generally reliable sources confirmed Iran's stealth fighter development issues and stagnant efforts. The reliability of sources was above average. All sources collaborated in favor of the analysis and mutually supporting that Iran's efforts to acquire and employ stealth fighter technology will lag in advancements given the availability of stealth fighter technology.

Authored by: Rafael Duran

[25] "Iran claims to have flown reverse-engineered US stealth UAV" London, Jane's Defence Weekly 10 Nov 14 (accessed through Jane's Defence Weekly): "Iran has flown a reverse-engineered copy of the Lockheed Martin RQ-170 Sentinel unmanned aerial vehicle (UAV) it captured from the United States in 2011, state media reported on 10 November. The announcement that an Iranian version of the classified UAV..."

[26] "Iran claims to have flown reverse-engineered US stealth UAV" London, Jane's Defence Weekly 10 Nov 14 (accessed through Jane's Defence Weekly): "Though details surrounding the payload have not been publicly released, images of the aircraft operating out of Kandahar Air Base in Afghanistan suggest that it is a sensor turret shrouded by a 'stealth-treated' window fairing...

Within the Next 10-20 Years, the U.S. Army Will Likely Need to Acquire and Improve Sensors to Detect Iran's Doctrinal/Non-Technical, Physical, Digital, and Electronic Signatures

Executive Summary:

Within the next 10-20 years, the current U.S. Army doctrinal/non-technical, physical, digital, and electronic offensive and defensive sensors and systems will not be likely capable enough to recognize, detect, and target Iran's cyber, unmanned systems, and missiles capabilities. Due to ease of access, availability of technology in the black-markets, affordable technology, and pressure from regional powers, Iran will likely continue to develop and enhance Cyber, UAS, and missiles capabilities. Despite the current international embargo, it is likely that Iran will successfully continue to acquire the essential materials for modernization.

Discussion:

Figure 57, Military & Aerospace Electronics. Israel Sensors Capabilities

There is no question that Iran is likely focused in a future modernization model supporting a regional domination strategy aiming to achieve offensive and defensive superiority. See the previous Iran's Cyber, UAS, and ICBM articles on this report for top capabilities modernization. Iran is making every effort to obtain, develop, and employ cyber offensive and defensive technology and "in light of continued attacks against Iranian cyber infrastructure, in March 2012, Iran announced plans to strengthen its cyber power by establishing a Supreme Council of Cyberspace to defend the country against cyber-attacks." [27] According to Israeli news sources, the headquarters was established in December 2012 and "by May 2015 the head of the Civil Defence Organization, Brigadier General Gholam Reza Jalali, announced the establishment of cyber defence workgroups

---

[27] "Iran – Strategic Weapon Systems" London, Jane's Sentinel Security Assessment – The Gulf States 17 Mar 20 (accessed through Jane's Group UK Limited): "Iran's offensive activities in cyber space are frequently fronted by an organization known as the Iranian Cyber Army, thought to be allied to the IRGC. The organization forms part of a cyber division allocated within the Iranian military structure.

to co-ordinate measures for defending nuclear facilities from cyber-attacks and the establishment of the Cyber Defence Centre under the IRGC's aegis." [28] Figure 57 depicts the effectiveness of sensors and systems to detect, recognize, and targets against future emerging unmanned aerial signature threats, H The predominant signatures likely difficult to detect within cyber, UAS, and Missile capabilities are doctrinal/non-technical, physical, digital, and electronics.

Future U.S. Army sensors and systems must be able to detect Iran's cyber-attacks to safeguards sensitive cyber systems. The U.S. Department of Energy (DoE), Center for Ultra-Wide-Area Resilient Electric Energy Transmission Network (NSF) set out on determining the best future Cyber defensive sensors conducting an study with the objective "to detect, recognize, and localize (both temporally and spatially) attacks from multiple sources using data collected from the ultra-wide-area monitoring network". H DoE concluded that future sensor and systems to identify threat cyber signatures must develop wide-area measurement systems (W AMS), and frequency-based real-time line trip detections.

However, due to the current turn in the oil market and the offset of the COVID-19, Iran's modernization momentum is likely to experience severe obstacles and stagnation. [29] See the previous articles on this report on Iran's Cyber, UAS, and ICBM top capabilities modernization. Iranian UAVs and missile attack in 2019 against Saudi Arabia expertly avoided air attack and area denial (A2/AD) sensors and systems, demonstrating the current rendering U.S. defensive systems ineffective. M Future U.S. sensors will likely need to increase the area coverage and have the technology required to acquire changes in doctrinal/non-technical, physical, digital, and electronics. The employment of future sensors must account for offensive and defensive operations. Defensive sensors are those systems that protect the intrusion of sensors and systems capabilities. Regulus Cyber, a professional leading cyber protection expert, explained the criticality of the cyber protection sensors and systems, and considering the increase "with sensors being used in both manned and unmanned platforms, the need for reliable sensor protection is increasing across all sectors and industries. From drones to driverless cars and from shipping to aviation, many assets are left exposed and vulnerable to malicious sensor-based threats. New solutions and technologies to protect sensors must be integrated into existing and future platforms of mobile phones, cars, ships, airplanes, drones, robots, and even infrastructure." H

[28] "Iran – Strategic Weapon Systems" London, Jane's Sentinel Security Assessment – The Gulf States 17 Mar 20 (accessed through Jane's Group UK Limited): "Iran's offensive activities in cyber space are frequently fronted by an organization known as the Iranian Cyber Army, thought to be allied to the IRGC. The organization forms part of a cyber division allocated within the Iranian military structure.

[29] "Iran – Executive Summary", UK, Jane's Sentinel Security Assessment – The Gulf States 19 Mar 20 (accessed through Jane's Group UK Limited): "Iran's economy is under severe pressure, with oil export arrivals at less than 300,000 barrels per day (bpd) in February 2020, down from above an average 1.2 million bpd in the first quarter of 2019. GDP will probably contract by 6.5% in fiscal year 2019 and 1% in 2020, with a highly likely further downward revision due to Iran's COVID-19 outbreak. Inflation – at more than 35% y/y – is expected to continue."

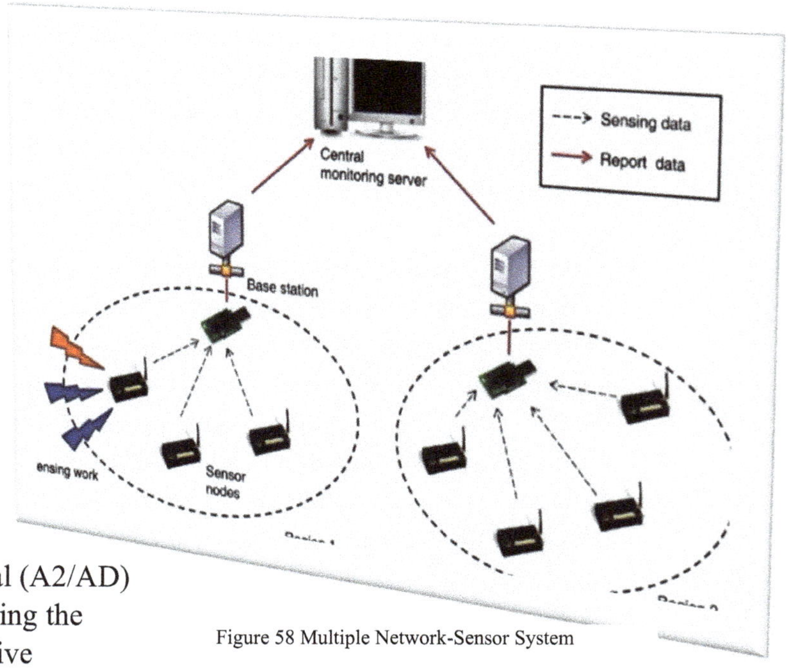

Figure 58 Multiple Network-Sensor System

However, due to the current turn in the oil market and the offset of the COVID-19, Iran's modernization momentum is likely to experience severe obstacles and stagnation. [30] See the previous articles on this publication on Iran's Cyber, UAS, and ICBM top capabilities modernization. Iranian UAVs and missile attack in 2019 against Saudi Arabia's U.S. issued air defense equipment, expertly avoided air attack and area denial (A2/AD) sensors and systems, demonstrating the current rendering of U.S. defensive systems as ineffective. M Future U.S. sensors will likely need to increase the area coverage and have the technology required to acquire changes in doctrinal/non-technical, physical, digital, and electronics. The employment of future sensors must account for offensive and defensive operations. Defensive sensors are those systems that protect the intrusion of sensors and systems capabilities. Regulus Cyber, a professional leading cyber protection expert, explained the criticality of the cyber protection sensors and systems, and considering the increase "with sensors being used in both manned and unmanned platforms, the need for reliable sensor protection is increasing across all sectors and industries. From drones to driverless cars and from shipping to aviation, many assets are left exposed and vulnerable to malicious, sensor-based threats. New solutions and technologies to protect sensors must be integrated into existing and future platforms of mobile phones, cars, ships, airplanes, drones, robots, and even infrastructure." H

The U.S. Army must consider the research and development (R&D) of future network intrusion detection systems (NIDS) for Iran's likely cyber and UAVs attacks against sensors. NIDS are sensors of their own that guard and monitor against attacks on other sensors. Traditionally, as sensors attempt to execute their intended purpose, NIDS either capture misuse detections or anomaly detections. Some advances NIDS can perform a combination of both protection operations, misuse and anomaly detections, but this type of protection is limited on wireless-type sensors. HThe vulnerability of sensors and

[30] "Iran – Executive Summary", UK, Jane's Sentinel Security Assessment – The Gulf States 19 Mar 20 (accessed through Jane's Group UK Limited): "Iran's economy is under severe pressure, with oil export arrivals at less than 300,000 barrels per day (bpd) in February 2020, down from above an average 1.2 million bpd in the first quarter of 2019. GDP will probably contract by 6.5% in fiscal year 2019 and 1% in 2020, with a highly likely further downward revision due to Iran's COVID-19 outbreak. Inflation – at more than 35% y/y – is expected to continue."

systems security in accordance with leading experts from the Molecular Diversity Preservation International (MDPI), including Ilkyu Kim, Doohwan Oh, et al, "sensor nodes in wireless sensor networks are easily exposed to open and unprotected regions. A security solution is strongly recommended to prevent networks against malicious attacks. Although many intrusion detection systems have been developed, most systems are difficult to implement for the sensor nodes owing to limited computation resources." [H] Figure 58 demonstrates the sensor composition and sensors defensive model to protect sensors from cyber-attacks on a multiple-network sensor system. [H]

Other defensive sensors are used to recognize and detect the presence of UAVs in a friendly area. These sensors' categories predominantly recognize and detect doctrinal/non-technical, electronics, thermal, heat, acoustic, and speed. The development of future sensors and systems to counter Iran's UAVs is a challenge because, according to HGH Infrared Systems, a global sensors and systems leading organization "UAVs small size and low electromagnetic signature go unregistered by traditional detection measures (radars); obtaining a drone is now possible for anybody, at a low cost; paraplanes, [used to smuggle drugs over the U.S. Southern border] are hard to spot and can carry hundreds of pounds of payload; and traditional detection measures are more susceptible to weather conditions". [H] As Iran continues to improve its UAVs to avoid detection, the employment of sensors must consider a combination of systems to detect, recognize, analyze, and target the future threat signatures. Thermal imaging technology

Figure 59 Electro-Optical Sensors, Intelligent Aero-Space Research and Development

and visible channel and laser ranger finders are part of sensors and systems that will enhance the U.S. Army's future capabilities to recognize, detect, and target UAVs. As per HGH Infrared Systems, a global provider of electro-optical devices for wide area surveillance and industrial thermography, "with thermal imaging technology, it is impossible for a drone to go unnoticed: any object, hot or cold will be detected by the

360° thermal sensor, day and night. Working in tandem with the Cyclope Intrusion Detection Software, Spynel tracks an unlimited number of targets (either airborne, terrestrial or maritime threats) to ensure that no event is missed over a long-range, wide area surrounding." [H]

Sensors and systems have demonstrated highest efficacy when utilized in a multimodal approach to best detect a myriad of signatures. In an MDPI Sensors Research, Samaras, Et al demonstrated that, "nowadays, c-UAV applications offer systems that comprise a multi-sensory arsenal often including electro-optical, thermal, acoustic, radar and radio frequency sensors, whose information can be fused to increase the confidence of threat's identification. Nevertheless, real-time surveillance is a cumbersome process, but it is absolutely essential to detect promptly the occurrence of adverse events or conditions." [H]

The U.S. Army will likely be more effective against Iran's and other adversaries' future UAVs, cyber, and missile attacks employing sensors and systems that, according to Samaras, Et al,"…include multiple integrated sensors for detecting the threat, mainly through radar and/or electro-optical/thermal (EO/IR) sensors and less commonly through acoustic and radio frequency (RF) sensors. Unfortunately, the majority of these systems are commercial applications…" [H] Figure 59, from a counter-UAV analysis, demonstrates the capabilities of each sensor and system when employed independently. [H] The employment of sensors and systems in a multimodal approach starts with the least range such as with the human eye to the highest range such as frequency detection and recognition. In terms of future materiel, the U.S. Army needs to emphasize on the R&D and the distribution to the individual level of UAVs detection optical-sensors to counter Iran's likely employment of micro UAVs, and the medium-sized unmanned combat aerial vehicle (UCAV).

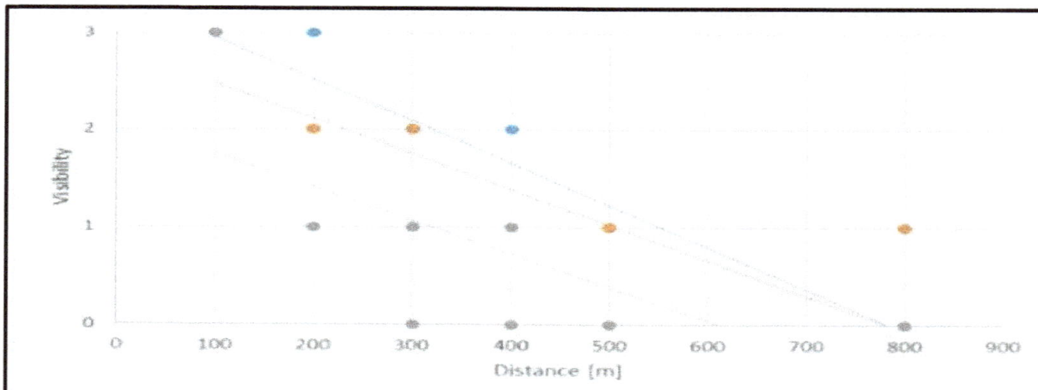

Fuzzy evaluation	3	2	1	0
Visibility	Good	Perceptible	Barely Perceptible	No visibility

Figure 60 Linear regression of the results of subjective visibility. (Multidisciplinary Digital Publishing Institute (MDPI)

Offensive future sensors and systems will allow the U.S. Army to effectively employ capabilities such as UAVs in a myriad of persistent reconnaissance and surveillance missions. These categories of UAVs mounted future sensors and systems mut be capable of numerous signature detections such as human intelligence (HUMINT), signal intelligence (SIGINT), electronic waves (EW) as well as the expected Iran's radio wave, thermal, infrared, and acoustic signatures. In the next 10 to 20 years, in order to effectively detect and acquire Iranian emerging signatures, the U.S. Army will likely need to improve wide-area electro-optical sensors mounted on UAVs facilitating the conduct of ISR operations. Future Army sensors and systems must provide "moderate-resolution persistent-surveillance electro-optical sensors that operate during day and at night over large areas to detect vehicles and people on foot." [H] Figure 59 from Intelligent Aerospace depicts UAVs sensors and systems mounted technology capable of identifying Iran's missile launching capabilities through the detection of Ultra-Wide-Area Resilient Electric Energy Transmission Network (NSF).

The least effective method of UAV signature detection is the human eye. Iran's future UAVs will likely increase in speed and lethality making the signature recognition though electronic means critical for defensive and offensive operations. The U.S. Army will require long range sensors capable of detecting UAVs threat thermal, acoustic, electric signatures at one thousand meters or greater to provide enough warning and weapon systems arming readiness. Experiments on the efficiency of the human eyes as a source of UAVs signature detection demonstrate that time, precision, and discrimination of UAVs is critical to safeguard personnel and facilities, and requires proper timely detection utilizing advanced technical instrumentation. Figure 60 illustrates the Basel, Switzerland sensors base analysis findings on subjective visibility which "concluded that the distance of effective UAV detection by human sight is about 200–300 m, which is very close to the IR detection limits when using commercial IR detection devices." [H]

Analytic Confidence:

Analytic confidence is moderate. The task to find information on Iran's sensors and system intention and progress technology was complex, and this estimate is sensitive to rapidly emergent information. The reliability of the sources was average. All sources corroborated supporting analysis on Iran's latest UAV demonstrations, sales, suspected attacks against Saudi Arabia, and the latest cyber-attacks reinforce the validity of sources.

Authored by: Rafael Duran

NORTH KOREA

North Korea's Estimative Key Findings

Authored By: Jerry Brown

Based on an analysis of North Korea modernization plans, defense spending and stated goals that evaluated 19 capabilities against 21 possible signatures, it is highly likely that four signatures—physical, visual, thermal signatures, and speed changes—will both impact the most capabilities and become more difficult, on average, to detect in the 2030-2040 time frame (See figure 61 below):

	Missiles	Long-Range Strike	Short-Range Strike	Stealth	Artificial Intelligence	Electronic Warfare	Unmanned Systems	Cyber	3D Printing	Air Defense	Anti-Satellite	Autonomous	Batteries	C4ISR	Chemical	Deception	Hypersonic	Quantum	Tanks
Physical	4	4	4	2	3	3	3	3	2	2	2	2	2	2	2	2	2	2	2
Visual	4	4	4	2	3	3	3	3	2	2	2	2	2	2	2	2	2	2	2
Speed	4	4	4	2	3	3	3	3	2	2	2	2	2	2	2	2	2	2	2
Electronic (Digital/Software)	3	3	3	2	3	3	3	3	2	2	2	2	2	2	2	2	2	2	2
Radio waves	3	3	3	2	3	3	3	3	2	2	2	2	2	2	2	2	2	2	2
Frequency	3	2	2	2	3	3	3	3	2	2	2	2	2	2	2	2	2	2	2
Acoustic	3	3	3	2	3	2	2	3	2	2	2	2	2	2	2	2	2	2	2
Thermal	4	3	3	2	2	2	2	3	2	2	2	2	2	2	2	2	2	2	2
Digital	3	3	3	2	2	2	2	3	2	2	2	2	2	2	2	2	2	2	2
Wavelength	2	2	2	2	3	3	3	3	2	2	2	2	2	2	2	2	2	2	2
Bio/Chemical	3	3	3	2	2	2	2	3	2	2	2	2	2	2	2	2	2	2	2
Infrared	3	3	3	2	2	2	2	3	2	2	2	2	2	2	2	2	2	2	2
Laser	3	3	3	2	2	2	2	3	2	2	2	2	2	2	2	2	2	2	2
Doctrine and other Non-Technical Process	3	3	3	2	2	2	2	3	2	2	2	2	2	2	2	2	2	2	2
Radiation	3	3	3	2	2	2	2	3	2	2	2	2	2	2	2	2	2	2	2
Seismic	3	3	3	2	2	2	2	3	2	2	2	2	2	2	2	2	2	2	2
Smell	3	3	3	2	2	2	2	3	2	2	2	2	2	2	2	2	2	2	2
Metanalytic	3	3	3	2	2	2	2	3	2	2	2	2	2	2	2	2	2	2	2
Ultra Violet	3	3	3	2	2	2	2	3	2	2	2	2	2	2	2	2	2	2	2
Hyperspectral	2	3	3	2	2	2	2	3	2	2	2	2	2	2	2	2	2	2	2
Magnetic	2	3	3	2	2	2	2	3	2	2	2	2	2	2	2	2	2	2	2
x-Ray	2	3	3	2	2	2	2	3	2	2	2	2	2	2	2	2	2	2	2

5 Much More Difficult to Detect	4 More Difficult to Detect	3 Moderate to Routine Improvement	2 Easier to Detect	1 Extremely Easier to Detect

Figure 61 North Korea and Capabilities Matrix Focus on Overall Signatures and Projected Capabilities

In fact, over 2% (10/418) of future North Korea threat signatures distributed over 4 of the 19 capabilities will likely become more difficult to detect between 2030 and 2040. 23% (97/418) will become moderate to routine to detect and 75% (313/418) of the signatures associated with these 19 capabilities are likely to become easier to detect in the same period.

	Short-Range Strike	Missiles	Long-Range Strike	Artificial Intelligence	Electronic Warfare	Unmanned Systems	Cyber
Physical	4	4	4	3	3	3	3
Visual	4	4	4	3	3	3	3
Speed	4	4	4	3	3	3	3
Electronic (Digital/Software)	3	3	3	3	3	3	3
Radio waves	3	3	3	3	3	3	3
Frequency	2	3	2	3	3	3	3
Digital	3	3	3	2	2	2	3
Thermal	3	4	3	2	2	2	3
Wavelength	3	2	2	3	3	3	3
Acoustic	3	3	3	2	2	2	3
Bio/Chemical	3	3	3	2	2	2	3
Infrared	3	3	3	2	2	2	3
Laser	3	3	3	2	2	2	3
Doctrine and other Non-Technical Process	3	3	3	2	2	2	3
Radiation	3	3	3	2	2	2	3
Seismic	3	3	3	2	2	2	3
Smell	3	3	3	2	2	2	3
Metanalytic	3	3	3	2	2	2	3
Ultra Violet	3	3	3	2	2	2	3
Hyperspectral	3	2	3	2	2	2	3
Magnetic	3	2	3	2	2	2	3
x-Ray	3	2	3	2	2	2	3

5 Much More Difficult to Detect	4 More Difficult to Detect	3 Moderate to Routine Improvement	2 Easier to Detect	1 Extremely Easier to Detect

Figure 62 North Korea Signature and Capabilities Matrix Focus on Top Signature and Projected Capabilities

1. Driving these changes are a number of key facts. Specifically:
    a. Economic sanctions by the United Nations and the United States.
    b. North Korea has relatively few technical experts.
    c. Journalist Barbara Demick, author of *Nothing to Envy: Ordinary Lives of in North Korea stated that* even without the sanctions, North Korea is a poor country with a limited budget available for modernization.

2. capabilities are likely to see the most upgrades, and, consequently, the most negative overall change (i.e., more difficult to detect) in 3 of the 22 signatures (Long- and

Short-Range Ballistic Missile, intercontinental Ballistic Missile) profiles over the next 20 years.

a. According to South Korean analysts, the North scraped together what little foreign exchange it had to buy $65 million of weapons from China and Russia. North Korea will likely continue to purchase equipment from China, Russia, or Iran and make moderate changes to EW equipment.
b. According to 38 North, Russia, China, and Iran funnel equipment to North Korea to support their ICBMs.
c. Based on Rand Arroyo Center, *Four Problems on the Korean Peninsula* Long-Range Strike and Short-Range Strike are North Korea's primary response to regional threat to the country.

3. North Korea is likely by 2040 to modernize long range and short range (LRBM/SRBM) strike capability but with nominal heat signatures changes.
    a. Long Range and Short-Range missiles come from Russia (9K720 and SS-26) with signatures already identifiable by U.S.
    b. North Korea funding for their ICBM is supported through their Cyber Warfare program. The U.S. Treasury Department said Friday that North Korean state-sponsored hacking groups attacked critical infrastructure, drawing illicit funds that ultimately funded the country's weapons and missile programs.
    c. Physical, Speed, and Visual will be difficult to identify by 2040 and will likely continue their research and development, albeit at a slow pace, to enhance by nominal changes in the signatures.

4. Korea unlikely to modernize GPS jamming signatures radio waves, frequency, wavelength for their EW program in 10 years.
    a. The economy of North Korea is severely repressed and has been the lowest ranked in the world since the inception of the Index of Economic freedom in 1995. Because of that North Korea realize on assistance from their allies to modernize their EW program.
    b. Due to limited economic capability and technological expertise. The majority of equipment purchased is older and those signatures have been identified. It is likely any modification to the purchased equipment would be nominal.

5. North Korea will likely increase their cyber capability by reducing malware tracking, digital imprints, and firewall manipulation by 2030.
    a. According to Public Radio International, generally, the country do not obscure its operations or digital imprint with cyber-attacks.
    b. North Kore is likely to evolve or adapt, improve cyber capabiltiies by increasing its malware, firewalls and hacking techniques.
    c. Based on Homeland Security Cyber-Infrastructure Security Agency North Korea improvement of cyber signature will likely develop as internet

capability matures but that the development will be capability matures but that development will be nominal.

6. Korea's ICBM missile programs is likely to result in making thermal, speed, physical and visual signatures more difficult to detect by 2035.
    a. Based on 38 North, the country has four ICBMs capable of reaching a US territory or military base. These include: the Taepodong-2, Musudan, KN-14, and Hwasong 15, depicted in in figure 63. Roh Jae-cheon, a spoken for the South Korean military Joint Chief of Staff stated that each year, these signatures are physically larger and present a larger thermal signature with two additional engine stages for longer reach capability.

Figure 63 Estimated Signature and Capability no-noise

    b. Brian Padden a Senior Correspondent for Voce of America, stated that North Korea mitigates UN sanctions by leveraging allies and black markets is becoming the new normal to obtain necessary materials for military purchase.
    c. By 2035, the next ICBM will likely have a <u>three-stage rocket system</u> that will have routine to moderate signatures (speed,

Figure 64 Hwasong-12, 14 and 15 Range Capability

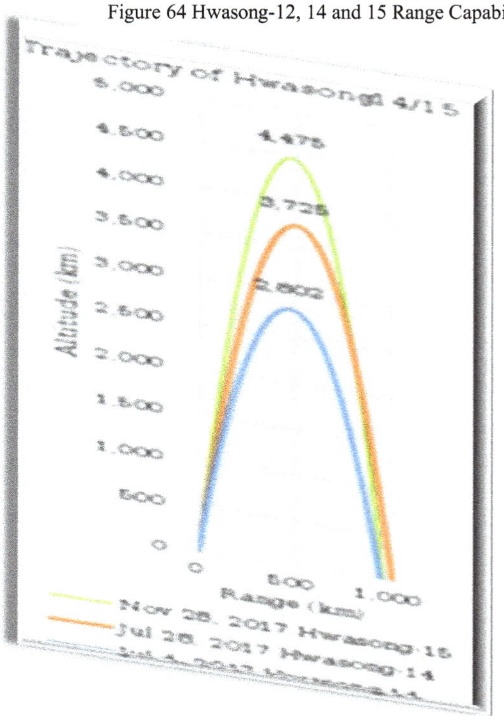

thermal and physical) changes comparable to the Hwasong-15 which is a two chambered main engine system as opposed to the Hwasong-12 or 14 which have on fixed main chamber and four gimbaled steering Vernier thruster chambers depicted in figure 64.

7. Removing capabilities which have nominal signature changes from the calculations, as seen in Figure 64, will create a slight increase in the percentage of signatures that become more difficult to detect. This may indicate that the U.S. Army will likely need Measurement and Signature Intelligence (MASINT) sensors and systems that can detect, track, identify or describe the distinctive characteristics of North Korea's visual, thermal, frequency, radio waves, and speed signatures. That broad signature approach will capture the key capabilities of the country's Artificial Intelligence (AI), Electronic Warfare (EW), Long-Range and Short-Range Strikes (LRBM/SRBM), Missiles Intercontinental Ballistic Missile (ICBM), Cyber and Unmanned Vehicle Systems (UAV).

8. North Korea has limited capability to mask its threat signatures. The US cannot easily detect all of NK signatures. North Korea does place limited emphasis on deception. Their courntry does place importance on developing technologies that will likely diminish their threat signature.

North Korea Likely to Enhance Cyber Capability to Sustain Nuclear, Intercontinental Ballistic Missile Programs

Executive Summary:

Due to North Korea's successful cyber-attacks, which have stolen 2 billion of dollars from Banks and cryptocurrency exchanges according to a UN report seen by seen by Reuters, it is likely that North Korea will invest significantly in the cyber program to supplement national funding for their purchasing of military equipment. Despite North Korea allocating 60% of its $30B (2019) Gross Domestic Product (GDP), for the military and military functions, North Korea depends on allies and the black market to sustain their nuclear and ICBM program.

Discussion:

According to the Freedome Natoinsl Index North Kroea has the second lowest GDP, $30B in 2019, North Korea continues to fund their missile program either through the blackmarket or from their allies such as China, Russia and Pakistan. North Korea is likely to enhance their cyber program to supplement funding to enhance their ICBM and Nuclear progam. A writer for the Public Radio Investigation (PRI) stated that "in 2018, North Korea developed the Reconnaissance General Bureau (RGB), North Korea's equivalent to the CIA, [and] has trained the world's greatest bank-robbing teams .[M] In

Figure 65 Unit 586 Reconnaissance general Bureau Building (Photo: KCNA) [M]

just the past few years, RGB hackers have struck more than 100 banks and cryptocurrency exchanges around the world, pilfering more than $650 million (Figure 65). The United States considers North Korea as one of the hardest countries to monitor and detect cyberattack activities. Overall, North Korea has generated an estimated $2 billion for its weapons of mass destruction programs using "widespread and increasingly sophisticated" cyberattacks to steal from banks and cryptocurrency exchanges. [M] In 2018, just after a summit meeting with the President of the United States, North Korea launched a cyberattack against South Korea stealing $32M. [M]

Funds from continued cyberattacks provides North Korea's regime with multiple options to fund the purchasing of parts for their nuclear weapons in the next 10 years. North Korea does not have to solely rely on their GDP when their cyber activity is used to supplement and support its priority programs. Despite the embargos placed on North Korea by the United Nations, analysis indicates that the North Korean Nuclear and ICBM program will be one of the most dangerous programs. However, in 2030-2040 North Korea is unlikely to be as dominant as the as current world leaders—the U.S., Russia, and China. Jeffrey Lewis, from Middlebury Institute of Strategic Studies stated, "The world will have to learn to live with North Korea's capability to target the United States with nuclear weapons". [M]

Analytic Confidence:

Analytic confidence in this estimate is *moderate*. Sources are reliable and generally corroborate each other. The analyst lacked experience but did use a structured method. There was adequate time but worked alone. While the analyst relied on open source for this information, this report is sensitive to change due to new information.

Authored by: Jerry Brown

By 2040 North Korea Likely to Modernize Long and Short-Range Strike Capability but with Nominal Heat Signature Changes

Executive Summary:

Despite economic sanctions on North Korea, they will modernize long and short strike missile capability by 2040 with nominal change to heat signatures. Due to North Korea's ally Russia, the long and short-range missiles recently fired by North Korea are reproductions of Russia's Iskandar Missile Systems, otherwise known as the 9K720 or the SS-26 that North Korea slightly modified.

Discussion:

It is likely that North Korea will modernize their long and short ballistic missile strike capability by modifying their launch thrust ratio that will enhance their speed, and increasing their thermal signature making these signatures more difficult to detect by 2030. (Figure 66) [M] Kim Jung-Un limits Korea's funds that are not dedicated toward his

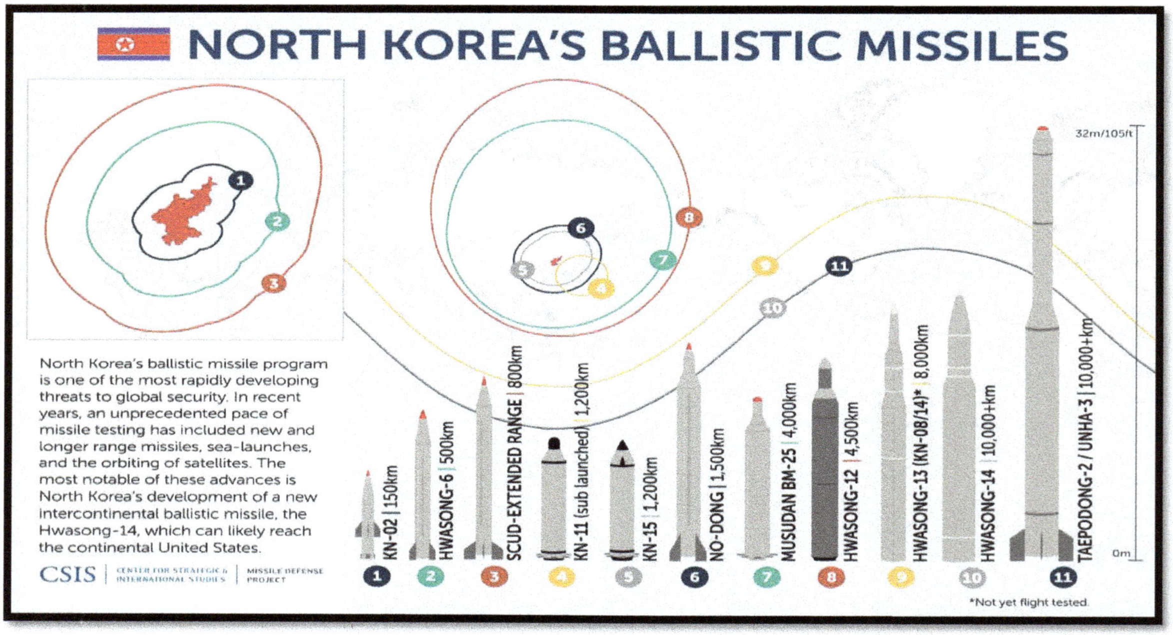

Figure 66: North Korea's intercontinental ballistic missile program [M]

missile program due to economic challenges. Instead the country are likely to modernize the entire missiles program by purchasing older long and short-range ballistic missiles (LR/SRBM) from Iran, Russia or China to enhance their capability. [L] North Korea's long and short strikes will have signatures that support electro-optical terminal guidance system which can be controlled from land or by aircraft. Their long and short-range strike signature will have the capability to modify the trajectory in flight which is the same signature as the Iskandar, capable of hitting moving targets as well as fixed sites and locations.

Although North Korea would like to develop their own LR/SRBM once their economic sanctions have changed, most of the current and future missile signatures will emulate the same missile technology from North Korean supporting countries such as China, Iran, and Russia. Russia, China, Iran and the Black market illegally deliver ballistic missile and other materials necessary for North Korean expansion capabilities from long- and short-range strike missiles. [M] North Korean allies have been an integral contributor to their LR/SRBM research and development (R&D) enhancing range capabilities and precision fires. It is likely that if North Korea conducts future long and short-range launches, the heat signature will have a nominal change based on the mission of their LR/SRBM which is to deter adversaries from attacking North Korea.

Despite North Korea's successful launched space satellites that demonstrated their countries technological capabilities which is required as they plan to modify their long and short-range strikes ballistic missiles by 2030. [M] If North Korea is capable of placing additional satellites in low orbit this would increase strategic defensive and offensive capabilities, and the ability to interrupt satellite communications, but the satellite launching platform signature will likely mirror the LR/SR launch.

Analytic Confidence:

Analytic confidence in this estimate is *moderate*. Sources were generally reliable and tended to corroborate one another. There was adequate time, but the analysis worked alone and was not a subject matter expert but used a structured analytic technique and was restricted to open source information.

Authored by: Jerry Brown

North Koreas Unmanned Aerial Vehicles Likely to Exhibit Nominal Heat and Digital Signatures Changes in the Next 10 Years

Executive Summary:

Due to a lack of technological expertise, North Korean Unmanned Aerial Vehicles (UAV)'s will likely exhibit nominal heat and digital signature changes in the next 10 years. Despite North Korean initiative by actively seeking to improve their drone capability, several crashed UAVs have been recovered by South Korea and were discovered to be primitive.

Discussion:

It is likely that North Korean UAV heat and digital signature will change at an incremental rather than dramatic pace by 2030. It is likely North Korean EW/UAV signatures will be similar in their thermal, speed, visual, electronics (digital/ software) indicating that in 2030 the US will likely detect similar UAV signature as they do now. [L] Based on an article from Korea JoongAng Daily, North Korea has been actively developing drones for military use, most notably deploying them to spy on South Korean military installations. However, in 2014 several lower quality UAVs crashed and were discovered by South Korea within their border (Figure 67). [M] While the UAVs in question appeared primitive, the South's inability to detect them rang alarm bells and generated significant controversy. [M] North Korea's ability to develop a different signature UAV that would challenge the US abilities to track and identity the threat depends on North Korea's ability to acquire new technologies from China, Iran or elsewhere. The cost of producing a UAV's ranges from $100,000 to hundreds of millions of dollars. [M] The Gross Domestic Product (GDP) in North Korea was approximately 18 billion US dollars in 2019. [H] According to Statista Research Department, statistic shows that an estimate of drone spending worldwide, from 2017 to

Figure 67 Suspected N. Korean drone spied on THAAD site: S. Korean military [M]

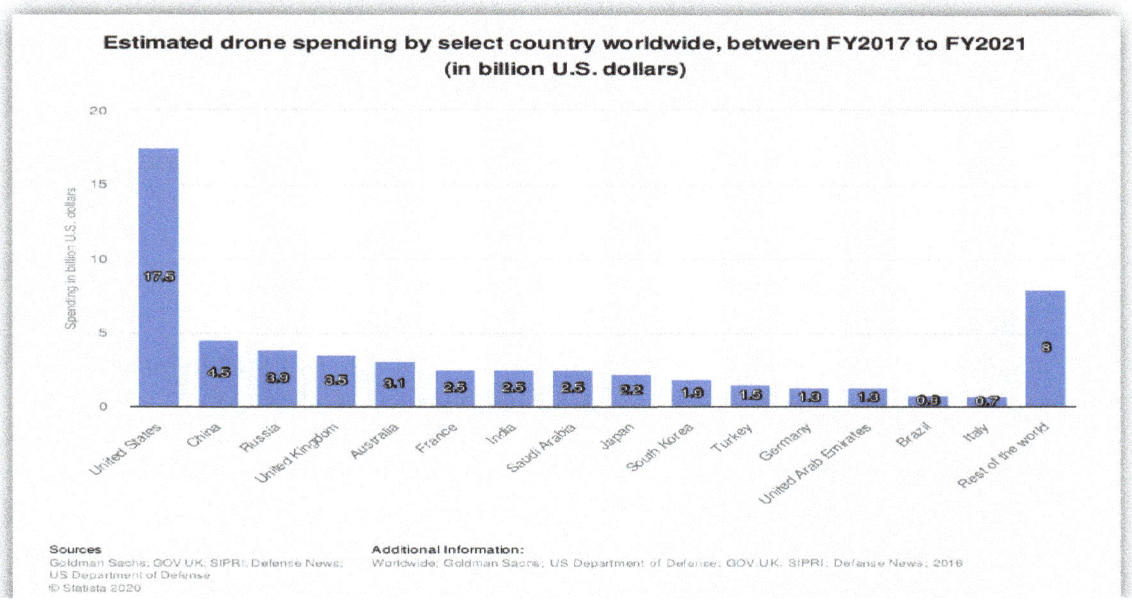

Figure 68 estimate of drone spending worldwide, from the 2017 fiscal year to fiscal year 2021 M

fiscal year 2021, by select country US is projected to spend about 17.5 billion dollars while North Korea in comparison is not even in the top 15 for estimated drone spending (Figure 68).

A few militaries, such as US, China, Israel, and Russia are already using UAV's for deterrent combat operations and reconnaissance programs. China recently finished building their CH7 which is jam-resistant and counter-stealth with quantum-radar platform. Due to North Koreas cyberattack activity, funding to support this advanced UAV can be generated. Stated earlier in the leaf regarding cyberattacks North Korea at one point netted more than $500 million per year from their cyberattack activities. M In 2018 just after a summit meeting with the President of the United States, North Korea launched a cyberattack against South Korea stealing approximately $32M. M China has surpassed the United States in UAV technology, thereby making North Korea a stronger adversary in the UAV platform, given its close relationship with China. M North Korea is not considered an expert in this field and appears to rely on China, Russia, Pakistan, and the black market to obtain high end UAV technology. H However, it is highly unlikely that North Korea would spend billions of dollars on a UAV program but is likely to spend billions on their nuclear and ICBM program. M

Analytic Confidence:

Analytic confidence in this estimate is m*oderate*. Sources were generally reliable and tend to corroborate one another. The analyst worked alone is not an expert. Furthermore, given the short time frame of the estimate, this report is sensitive to change due to the new information.

Authored by: Jerry Brown

North Koreas likely to Increase Their Cyber Signature by Reducing Malware Tracking, Digital Imprints, and Firewall Manipulation by 2030

Executive Summary:

Despite North Korea's gross domestic product (GDP) $30 billion, the country is evolving their techniques of malware deception, digital imprints, and firmware by leveraging their cyber program. North Korea will likely evolve their cyber capabilities and reduce their malware tracking signature by 2030, making it more challenging to detect.

Discussion:

It is likely that North Korean cyber technology will evolve by 2030 in two ways. First, it will evolve through the use of already developed industry malware, firewalls, and hacking techniques. [M] Secondly, the speed at which they develop this technology will improve as North Korea continues to revolutionize their cyber program with faster internet service. [H]

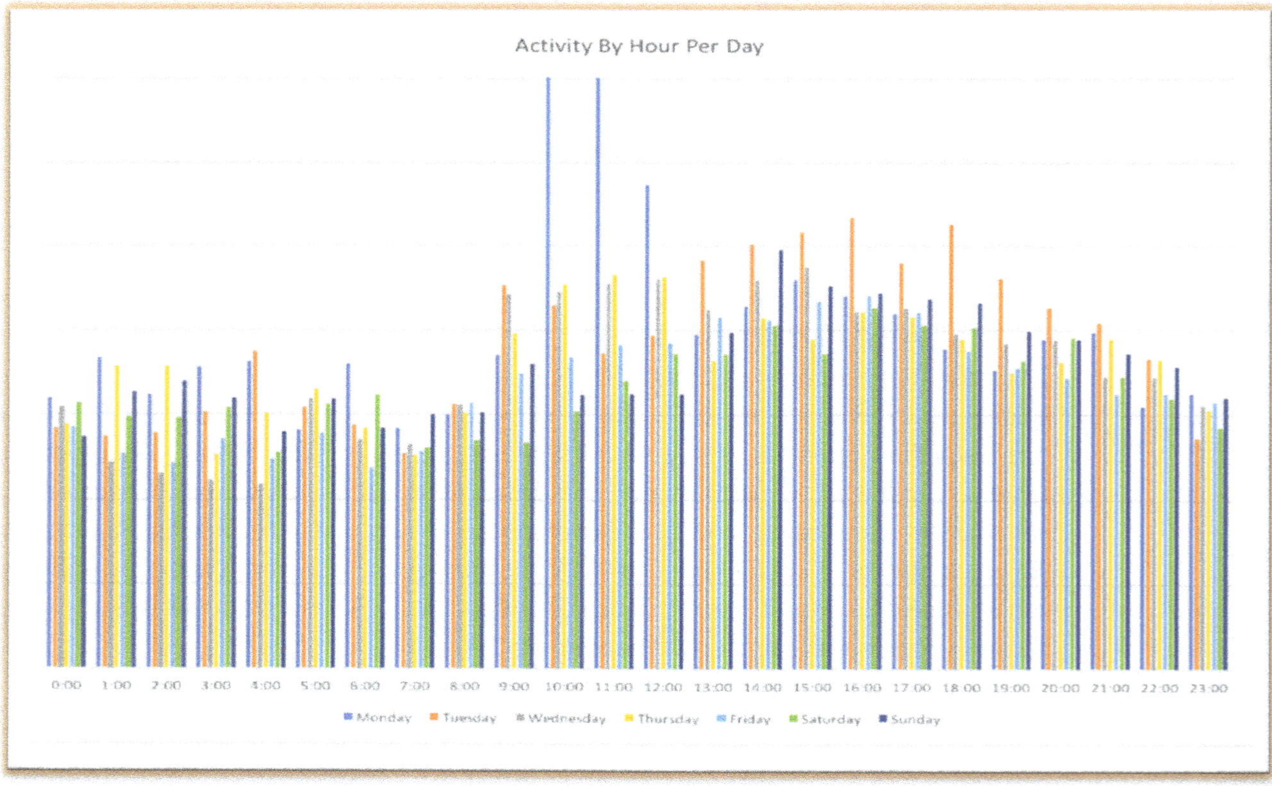

Figure 69 North Koreans' distinct patterns of daily usage [M]

North Koreans have distinct patterns of daily usage over a specific period (figure 69) as well. M During the weekdays, times of highest activity are from approximately 9:00 a.m. through 8:00 p.m. or 9:00 p.m. local time, with Mondays and Tuesdays consistently having the highest activity (Figure 69). Historically, North Korea conducts cyberattacks from already known and well-understood technical networks. H To be sure, the networks able to support large-scale cyberattacks are frequently difficult to build and maintain. They use masked approaches, which mitigate the risk of direct attribution, by launching an attack from an identifiable home base. Once established on a network, however, North Korean cyber operations have an identifiable signature from the implants they use, therefore, they lessen the difficulty for the US to determine who is behind the hack. For example, the North Korean regime has used the same wipers and ransomware in the following attacks: WannaCry, Sony Pictures, and the Taiwanese Bank. M

Despite North Korea's efforts in evolving their cyber signatures, the US can follow the movement of North Korean cyber hackers via the Cybersecurity and Infrastructure Security Agency (CISA), the Federal Bureau of Investigation (FBI), and the Department of Defense (DoD). These entities have known that North Korea uses the malware files HopLight, BistroMath, SlickShoes, HotCroissant, ArtfulPie, BuffetLine, and CrowdedFlounder. H North Korea, as mentioned, uses malware to reduce their digital imprint and signature. BuffetLine, particularly, appears to encrypt its traffic in a way that fakes Transport Layer Security (TLS) encryption, which can make nefarious activity blend into normal traffic. BuffetLine is also capable of manipulating file timestamps, so the North Korean hackers can obfuscate their activities to an extent. M

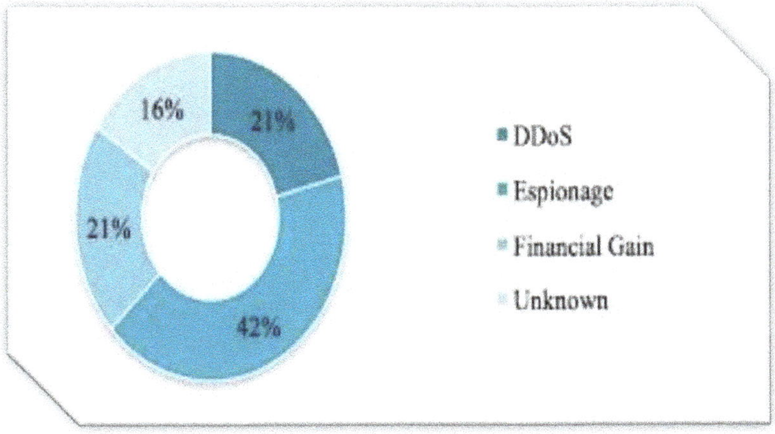

Figure 70 Percentage of North Korean cyberattacks against South Korea by type, with Espionage attacks representing the largest proportion M

North Korea's cyber program continues to evolve as a threat to the US and will create a greater challenge to track by 2030. North Korea's cyber program has proven they are capable of penetrating the US cyber program, creating mayhem wherever North Korea decides to attack. M Figure 70 illustrates the distribution of North Korean cyberattacks against South Korea, another adversary, by type. North Korea's Kim Jong-Un understands the importance of cyber warfare within asymmetric capabilities and has gradually developed its cyber power in sequential phases. North Korea may be politically isolated, but the country is suspected of having thousands of hackers highly capable of carrying out global cyberattacks, like the recent ransomware attack in more than 150 countries. [H] In February 2020, North Korea expanded their targets to companies in Turkey, operating from a block of internet addresses traced to Namibia, one of the few countries that still maintains friendly relations with Pyongyang. [M]

Analytic Confidence:

Analytic confidence in this estimate is moderate. Sources are reliable and generally corroborate each other. The analyst lacked experience but used a structured method. There was adequate time but worked alone. While the analyst relied on open sources for this information, this report is sensitive to change due to new information.

Authored by: Jerry Brown

North Korea Unlikely to Modernize GPS Jamming Signature Radio Waves, Frequency, and Wavelength for Their EW Program in 10 Years

Executive Summary:

Due to limited economic capability and technological expertise, it is unlikely that in 10 years North Korea will modernize their Global Positioning System (GPS) Jamming capability using radio waves, frequency and wavelength. North Korea's current jamming capabilities have a nominal effect on the United States (US) and will likely purchase older equipment form China and Russia, or through the black-market in order to modernize their GPS Jamming signature capabilities.

Discussion:

Due to North Korea's limited economic capabilities and, their $30B gross domestic product of which 60% is spent toward their Nuclear and Intercontinental Ballistic Missile (ICBM) program, it is highly unlikely they will spend additional funds to improve their GPS jamming capability. M North Korea's primary purpose for purchasing GPS mounted jamming devices from Russia and China were to deter South Koreans use of GPS and to resemble Russia (Figure 71). M The technology in these truck mounted GPS jammers is old, however, North Korea successfully disrupted South Korea's military operations in 2012. The US has countered their older technology to prevent North Korea from interfering with South Korea's military operations. M Therefore, it is unlikely that there will be significant signature changes (radio waves frequency, thermal) within the next 10 years that have not been already identified by the US. The power amplification signatures of North Korea would likely resemble Russia's and China's (Figure 71), that is continuous noise/masking jamming from 1227.6MHZ;1575.42 and 1500to 1900 MHZ. M The director of the Defense Intelligence

Figure 71 Russian R-300ZH Zhitel Jamming Cellular Satellite communication Station System M

Agency (DIA) LTG Robert Ashley noted that "North Korea is a low-level threat to the United States in Northeast Asia and but is our hardest intelligence collection target due to how hard the information is to collect". [M] North Korea will likely continue to invest into their Electronic Warfare (EW) capability purchasing older equipment from Russia, China or the black market. Because of that, North Korea will likely modernize but it will be such a nominal effect on their EW signature in next 10 years. [M] For further information on Russia EW jamming capability review Russia EW report.

Analytic Confidence:

Analytic confidence is *moderate*. The sources were generally reliable and tended to corroborate one another. There was adequate time and use of the analytic technique. The analyst is not a subject matter expert and was restricted to using only open source information which was limited. Due to the merging and growing technology this report is time sensitive to change.

Authored by: Jerry Brown

North Korea ICBM missile programs is likely to result in making thermal, speed, physical and visual signatures likely easier to detect by 2035

Executive Summary:

Despite North Korean economic sanctions and a low Gross Domestic Product (GDP), North Korea will likely enhance their Intercontinental Ballistic Missile (ICBM) program by modernizing their thermal, speed, physical program likely easier to detect by 2035. Due to North Korea's relationship with Russia and China, these countries likely provide the materials to enhance North Korean programs. North Korea is also purchasing material through the black market limiting the effectiveness of United Nations sanctions.

Discussion:

It is likely that future North Korean ICBM and missile program will support a higher thermal signature with increased velocity (speed) from a possible multiple stage rocket

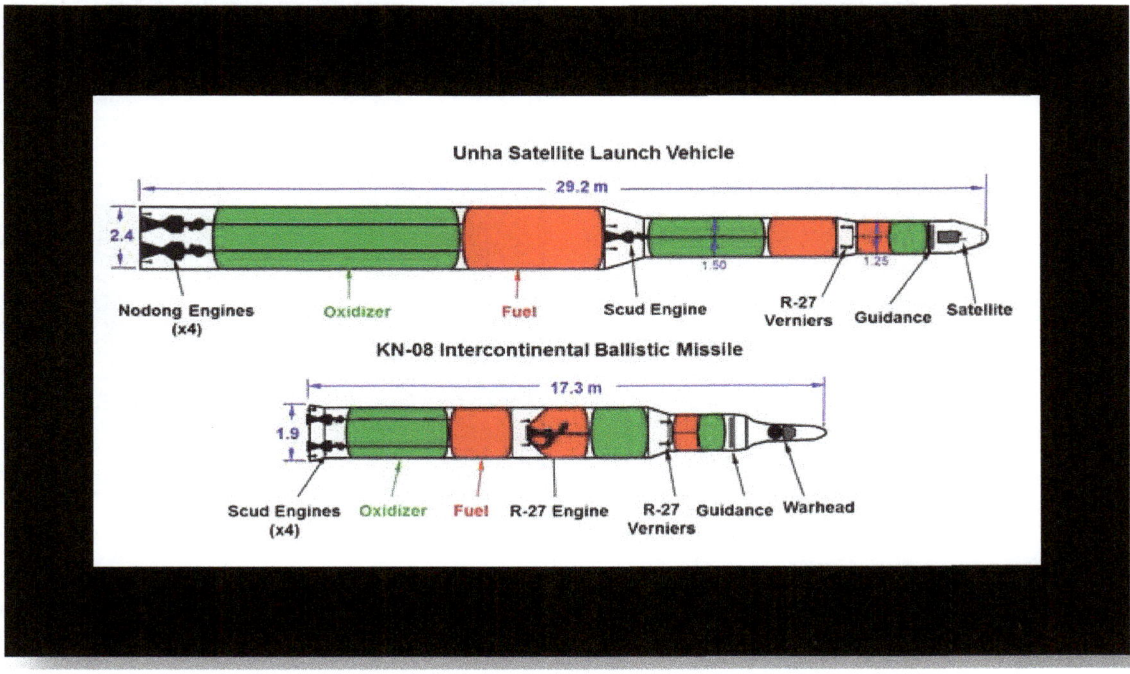

Figure 72 North Korea three-stage, liquid-fueled Unha rocket

engine that will shorten the trajectory and speed time, once launched. The launch would be similar to North Korea Kn-08 three stage satellite launching rocket (Figure 72). [H] North Korea's Hwasong-12 and the Hwasong-14 use four small engines mounted in parallel to the main thrust chamber for missile control during the first stage of the flight. North Koreas engineers have mounted each main engine of the Hwasong-15 on a slotted gimbal allowing each to be reoriented in one-dimension to control the direction of the exhaust gases, providing control. M North Koreans would venture out to purchase the stage boosters' power by an RD-107A engine that the Russians used on their Soyuz-21b rocket. (Figure 72) M

It is likely that funding for North Koreas modernized engines will come from their cyberattack activities (See NK Cyber report) M. Despite a very low GDP, $30B in 2019, North Korea continues to fund purchase parts for the ICBM program either through the black market or from their allies, China, Russia and Pakistan. A writer for the Global Post Investigation (GPI) stated that "in 2018, North Korea developed the Reconnaissance General Bureau (RGB), North Korea's equivalent to the Central [Intelligence Agency] CIA, [and] has trained the world's greatest bank-robbing teams". M In just the past few years, RGB hackers have struck more than 100 banks and cryptocurrency exchanges around the world, pilfering more than $650 million. M Overall, North Korea has generated an estimated $2 billion for its weapons of mass destruction programs using "widespread and increasingly sophisticated" cyberattacks to steal from banks and cryptocurrency exchanges. M In 2018, just after a summit meeting with the President of the United States, North Korea launched a cyberattack against South Korea stealing $32M. M

Figure 73 Stage III of the Soyuz rocket attached to its payload Russina Pace Lab

Funds from continued cyberattacks provide North Korea's regime with multiple options to grow its ICBM program by 2030. United Nations analysis indicate that North Korean ICBM and missile program will be one of the most challenging programs by 2030-2040. Jeffrey Lewis, from Middlebury Institute of Strategic Studies stated, "We're going to have to learn to live with North Korea's ability to target the United States with nuclear weapons." M

Analytic Confidence:

Analytic confidence in this estimate is *moderate*. Sources are reliable and generally corroborate each other. The analyst lacked experience but did use a structured method. There was adequate time but worked alone. While the analyst relied on open source for this information, this report is sensitive to change due to new information.

Authored by: Jerry Brown

ADDITIONAL INFORMATION

Technological Advances and Cost Reductions as well as Advances in Multi-Sensor Integration Will Enhance all Autonomous in the Next 10 – 20 Years

Executive Summary:

Due to continuous commercial developments in Artificial Intelligence (AI), Machine Learning (ML), technological innovations, cost reductions in sensors, as well as advances in multi-sensor integration; safe and automated motion will likely become a reality in the next 20 years. Despite several countries working towards enhancing autonomous machines there is a lagging AI open source community and technological innovation ecosystem. Many are pushing to make strides in autonomous development in both their commercial and military sectors.

Discussion:

The ensuing creation of supercomputers and upcoming developments in 5G networks are "important prerequisites for autonomous technologies that can process data in real time" enabling them to make split second decisions. [H] Advances in technology are surely to provide lower costs for various commercial and military industries. [H] "Cost reductions in sensors, actuators, radar, lidar and camera systems, as well as advances in multi-sensor integration through sensor fusion, improve depth detection for safe and automated motion and bring autonomous things closer to reality. . .[r]apid advances in areas such as AI, ML, and Deep Neural Networks are creating the conditions for autonomous navigating machines" [H]

Figure 74 Autonomous Vehicle Use in Wuhan China during Corona Virus Outbreak [H]

Click Picture to be connected to video (must have internet connection)

Various autonomous platforms such as vehicles, drones, and warehouse machines are constantly being improved or developed to be deployed to for use by the general public. As seen in recent commercials and leveraged by companies such as Amazon, automation in the logistic, warehousing, and delivery chain is becoming a competitive market. [H] "The market for mobile robots, drones, and autonomous vehicles in delivery and warehousing is likely to reach a staggering $81 and $290 Billion in 2030 and 2040, respectively." [H] These commercial enhancements have pressured various entities to attempt and pursue their own saving and competitive measures. As an example, in order for China to reduce its dependence on American chips, "the Chinese government has poured in huge amounts of resources to grow the domestic chip industry." [M] Chinese chip development will make it more difficult for American or Western detection of platforms with which they would be imbedded, even though Chinese companies such as Alibaba and Baidu have joined US companies such as "Nvidia, Intel, and Google to create chips purposely designed for AI," [M] which can form an Autonomous Vehicles (AV) nerve center.

While AV are a good test for Autonomous Systems, the AV future wave has yet to reach the fifth and final level of autonomy, on a zero to five scale. The final level is currently hindered because of "data storage size, data transportation, sensors costs, acquisition of corner case data, training data acquisition, and verifying deep neural networks." [H] This is most likely a similar reason lending to the slow innovation in various countries of autonomous systems.
Autonomous platforms and surveillance AI are being exported by China to countries in the Middle East and Africa. [M] "Lethal autonomous weapons would be commonplace. . . ever-increasing use of AI is inevitable. . .[and are] consistent with ongoing Chinese autonomous military vehicle development programs and Chinas current approach to exports of military unmanned systems." [M]

Analytic Confidence:

The analytic confidence for this estimate is *moderate*. Open sources were used and were generally reliable. Source content varied; however, sources tended to corroborate one another. There was adequate time allotted to the task and the analyst used a structured research method.

Authored by: Ashraf Abdelhak

Adversaries Are Unlikely to Use Artificial Humans via Artificial Intelligence to Spread Disinformation in the Next 10 Years

Executive Summary:

Programmability and the speed of life-like qualities that machine learning allows unlikely to give adversaries such as, Russia, China, Al Qaida and others an opportunity to leverage Artificial Humans (AH) or "virtual individuals" to perform complicated tasks on their own in the next 10 years. Artificial Intelligence (AI) has matured to a point where it can communicate and distribute roles among robots, as well as choose a commander. Given this development it may also be able to spread disinformation on the battlefield, but is unlikely to happen in the near future due to computer processing speeds. Despite being intended for commercial applications, defense ministries and defense departments worldwide could use AH to shape the battlefield and affect senior level decision making through impersonation.

Discussion:

Samsung Technology and Advanced Research Labs sub-company Neon debuted its AHs project at Consumer Electronics Show (CES) 2020. Neon said their design of AHs are programable and able to have real conversations and behave like humans. The AHs are able to "form memories and learn new skills, but don't have a physical embodiment." [H] Designers can personalize AH to be "goal-oriented" in completing tasks or personalized to assist as teachers, healthcare providers, spokespeople, or television anchors.

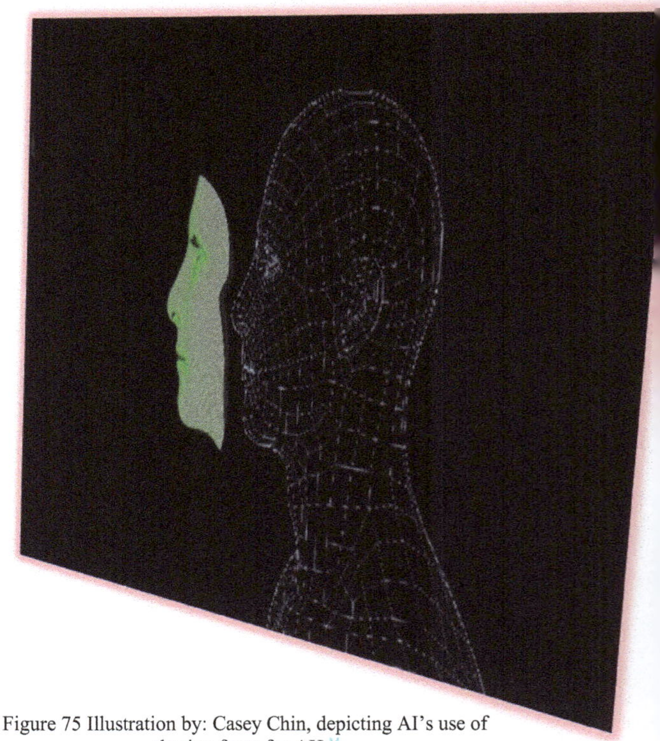

Figure 75 Illustration by: Casey Chin, depicting AI's use of altering faces for AH [M]

AHs development further enhances Russian owned United Instrument Manufacturing Corporation (UIMC) software package called Unicom. The Unicom "AI software gives military or civilian robots enough AI to perform complicated tasks completely on their own." [M] Processing power is needed to get these AI's to perform at the AH level, however UIMC's Deputy General Director Sergey Skokov said "this is yet another step to the creation of fully-fledged [AI], enabling mechanisms with nearly human capabilities." [M]

AHs are "lifelike, scalable, and virtual" individuals who can go anywhere a human can go as long as you have a platform to project the information, image, or at some point a hologram. [H] Wired Magazine contributor, Will Knight said, "Video manipulation and deception have long been possible, but advances in machine learning have made it easy to automatically capture a person's likeness and stitch it onto someone else," giving additional fidelity to the advent of AH and maturity of AI. [M]

Even though AI and AH are developing, companies such as Google and Facebook are overseeing projects focused on being able to spot AH videos or technology. [M] Developing the needed software to spot such programs should stay at the forefront and ahead of AH development. According to Neon's CEO, Prana Mistry the "main limitation right now is the requirement for a large amount of local processing power to render each avatar live." [H]

Analytic Confidence:

Analytic confidence on this estimate is *moderate*. The task to uncover adversarial development of AH on the battlefield was particularly complex. The reliability of the sources available on this topic were above average and the information contained was first-hand knowledge. The sources further tended to corroborate each other.

Authored by: Ashraf Abdelhak

Robots and Artificial Intelligence Likely to Enhance Developments of Biological Weapons via Additive Manufacturing in the Next 10 Years

Executive Summary:

Despite the fact that there have been decades of campaigning and arms control treaties set in place, the use of robots and Artificial Intelligence (AI) is likely to enhance biological weapon development in the next 10 years through Additive Manufacturing (AM), also known as 3D printing. Countries based in Europe, Africa, and other parts of the world tend to have the needed materials to test and produce these weapons. While such countries such as France and Germany who were rated in the top five exporters in addition to the United States (US), Russia, and China between 2014 and 2018 helped contribute to some of the largest weapons clients including South Korea, India, Israel, Egypt, Greece, and Saudi Arabia; setting a path for increased weapons proliferation.

Discussion:

Additive manufacturing, also known as 3D printing, AI, and robots are driving the development and use of a wide variety of weapons and are likely to have a significant impact on the development of biological weapons over the next 10 years. [H] According to Dr. Vincent Boulanin, Senior Researcher at the Stockholm International Peace Research Institute (SIPRI) on emerging technologies shares that "[t]he increased use of robots in laboratories could lead to significant gains in productivity during the design-build-test cycle of biological weapons, while artificial intelligence could be used to find new ways to optimize the transmissibility or virulence of a biological agent." [H]

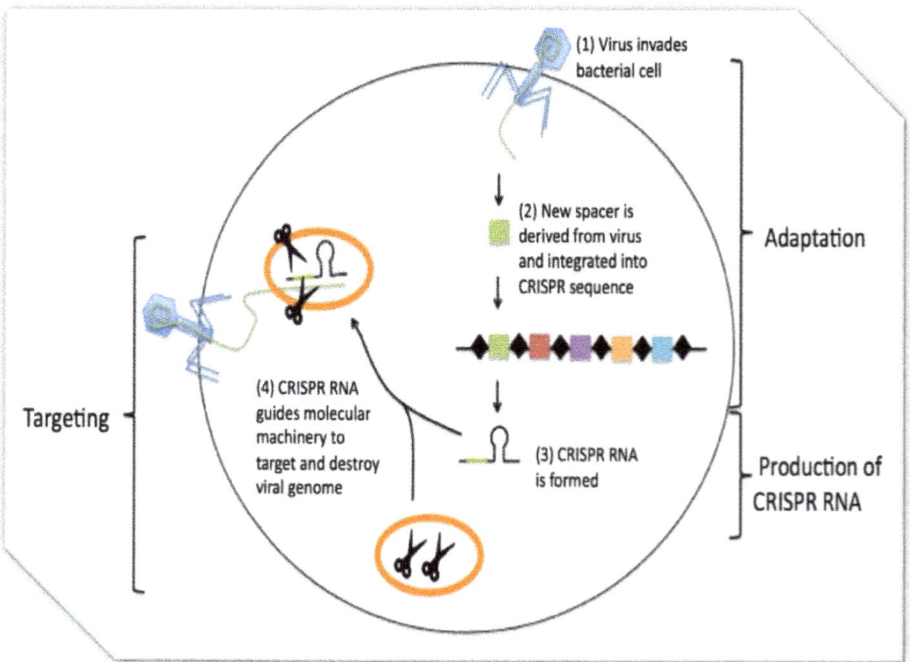

Figure 76 Illustration of CRISPR: A game-changing genetic engineering technique [H]

Online databases which contain genomic data and access to the internet in remote areas will provide immoral actors the needed tools to develop some level of needed weapons. "In the future, genomic data, gene editing tools such as [clustered regularly interspaced short palindromic repeats] CRISPR, and machine learning tools may assist nefarious actors interested in developing more effective biological weapons." [H] As future trends become more digitized, it will allow evil "actors to move fluidly between the digital and physical worlds, circumventing efforts of counter WMD proliferation in ways that the U.S. defense enterprise is not prepared to manage. . .[further leveraging,] physical-to-digital conversion technologies, such as gene sequencing and 3D printing, turn physical matter into digital information that computers can read, analyze, and share. " [H]

While industrial-scale 3D printers are already being used to advance technology and weapons testing, labs are also advancing science in the field. Robert Shaw, Middlebury

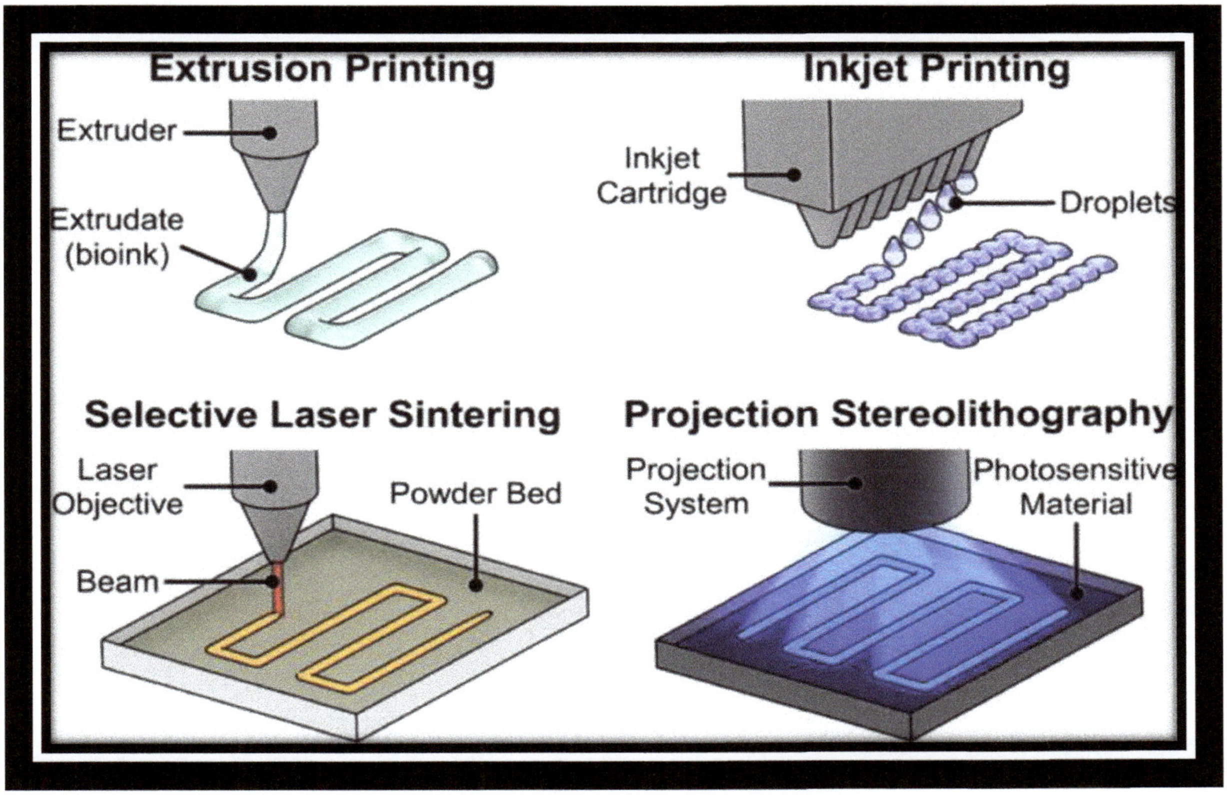

Figure 77 Selected additive manufacturing (AM) techniques [H]

Institute of International Studies at Monterey, Director of Export Control and Nonproliferation Program, shares the "proliferation of 3D printers, combined with advances in artificial intelligence, could make it much easier for nations or individuals to covertly build nuclear, chemical, and biological weapons." [M]

AM has been embraced by several countries, some of which are under sanctions or have the highest interest in disrupting the US' version of World Order. "AM technology and expertise can be found in many countries across the world. . . [such as] Germany, China, India, South Africa, Taiwan, and Iran." [M]

AI can definitely assist the printing process and prevent errors, as it is often related to expressions such as machine learning, neural networks, automation or artificial vision. When combined with 3D printing, it could increase performance and reduce the risk of errors by facilitating automatic production. [M]

Analytic Confidence:

Analytic confidence on this estimate is *moderate*. Open sources were used and were generally reliable. Source content varied; however, sources tended to corroborate one another. There was adequate time allotted to the task and the analyst used a structured research method.

Authored by: Ashraf Abdelhak

French and German Arms Trade Likely to Increase Over the Next 10 Years Despite Global Calls to Limit Proliferation

Executive Summary:

Despite the fact that there have been decades of campaigning and arms control treaties set in place, arms and weapons trade is likely to increase over the next 10 years. Countries such as France and Germany who were rated in the top five exporters in addition to the United States (US), Russia, and China between 2014 and 2018 helped contribute to some of the largest weapons clients including South Korea, India, Israel, Egypt, Greece, and Saudi Arabia; setting a path for increased weapons proliferation.

Discussion:

Arms sales are no new matter to the world stage as conflicts continue throughout the world and regimes attempt to reign control over their country masses. The global arms industry sales are on the rise. As of December 2019, sales are up 4.6 percent worldwide. [H] While the US dominates the world in arms sales for the first time since 2002, Russian companies' arms sales remain somewhat stable, only falling a fraction of a percent. [H] As depicted in Figure 78 Germany and France do have a large chunk of the world exports.

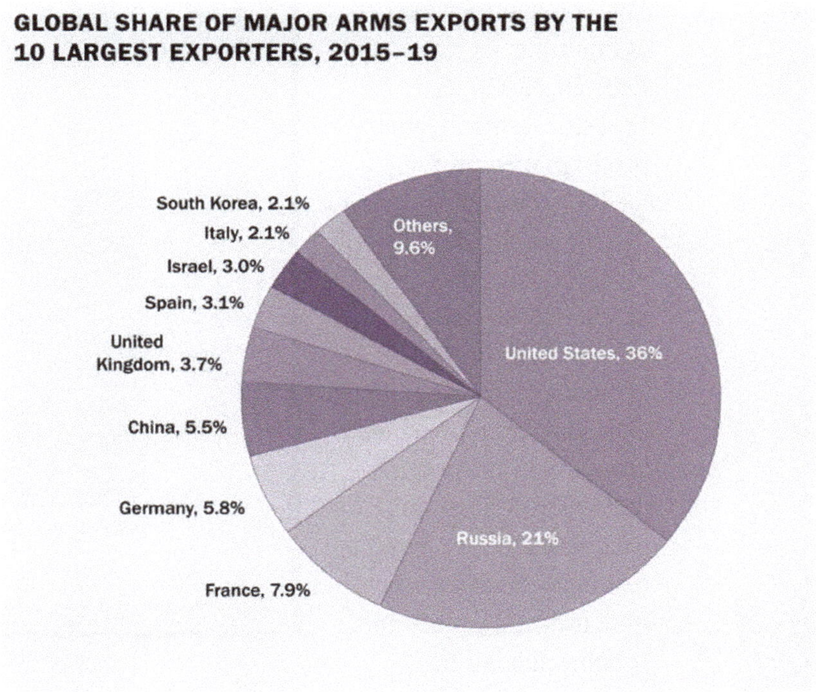

Figure 78 SIPRI Arms Transfer Database March, 2020 [H]

The combined arms sales of 27 European companies increased marginally in 2018, while combined sales of French companies rose 30% and German companies fell by 3.8%. [H] Due to global demands, for the same year, France saw an increase in their combat aircraft while Germany's decreased due to their low sales in ship production. French exports are assessed to continue increasing as they focus on robotic and artificial intelligence development given their armed forces announcement of increased spending in 2019. [H] As illustrated in July 2019, during Frances National Day, the parade of secrete weapons and new military equipment included Unmanned Ground Vehicles (UGV) which was developed by the Franco-German Research Institute and fitted with a navigation system to perform autonomous missions. [M] French are further developing exoskeleton armor which can be worn by humans, micro unmanned aerial vehicles developed by Prox Dynamics of Norway, which is leveraged by French Special Forces and is the world's smallest intelligence, Surveillance, and Reconnaissance (ISR) platform, and other technological systems. [M]

As of 2018 – 2019, France's recent jump and future military sales is due to its fighter jet exports to Egypt, India, and Qatar while its naval exports have been to Brazil, India, Malaysia, Belgium, Netherlands, and the United Arab Emirates (UAE). [H] Germany's solid sales efforts have been to a less wide audience, including European Member states, as well as volatile regimes such as Egypt and the UAE. [M]

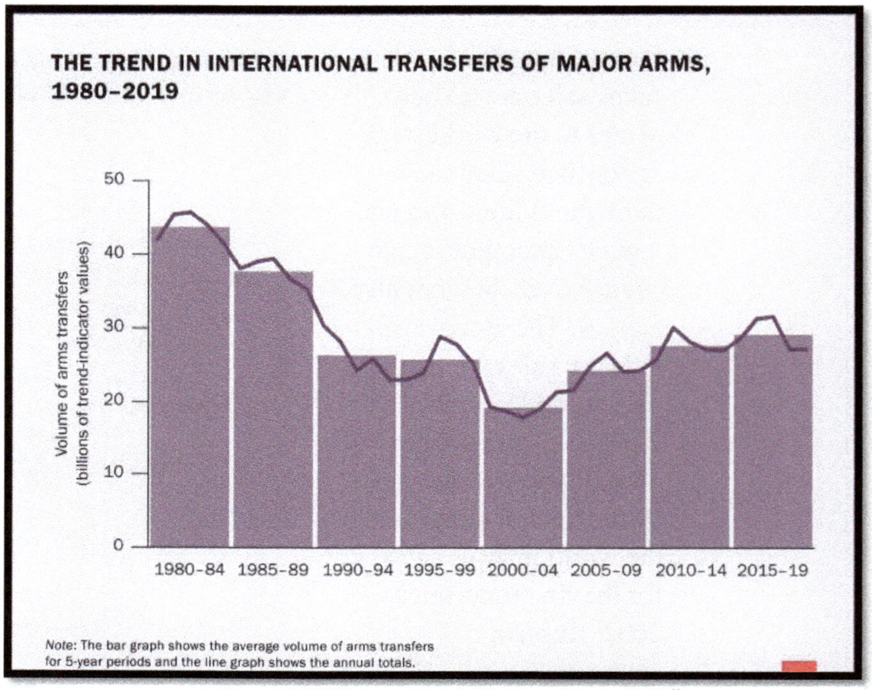

Figure 79 SIPRI Arms Transfer Database March, 2020 [H]

While not set for export at this time, as recent as February 2020, the French-German fighter program cleared for technology demos for their joint combat aircraft program with the possibility of Spain joining later in the program. "France and Germany each provided around $83 million for the new program stage. . . [which includes] Air Combat Cloud [that] will in real time connect and synchronize all the platforms and enable the processing and distribution of information to enhance situational awareness and collaborative operations." [H] The world tends to work in cycles as technology enhances the way of life and modernizes future weaponry, it would be a surprise to see an upward trend of future sales as depicted in figure 79.

Analytic Confidence:

Analytic confidence on this estimate is *moderate*. Open sources were used and were generally reliable. Source content varied; however, sources tended to corroborate one another. There was adequate time allotted to the task and the analyst used a structured research method.

Authored by: Ashraf Abdelhak

Spain's Global Arms Exports Likely to Increase in Next 10 Years; Comparable to Germany & France Arms Export Levels

Executive Summary:

Despite Spain's defense industry declining by 7.5% from the previous years, it is likely its arms exports will increase over the next 10 years. An increased arms purchases from France, Germany, United States (US), United Arab Emirates, Russia, and China will likely contribute to Spain achieving a larger global share of exports by 2030.

Discussion:

As of December 2019, arms sales are up 4.6 percent worldwide. [H] Spain ranks 7th in the global share of major arms exports (see figure 80). [H] Spain's arms exports, by volume, increased by 13% since 2010-2014. [H] According to Stockholm International Peace Research Institute (SIPRI), Spain's top arms customers are Australia, Singapore, Turkey, and Spain mostly buys arms from Australia, South Korea, Turkey, Singapore, Kazakhstan, and Mexico. [H] Figure 80 graphically depicts Spain's weapon export value and destinations and suggests a likely steady continued increase in sales.

In 2019, when the UN Security Council condemned foreign military involvement in Libya, the UAE had major arms import deals ongoing with Australia, Brazil, Canada, China, France, Russia, South Africa, Spain, Sweden, Turkey, the United Kingdom and the USA. [H]

According to data from the Spanish Defense Industry Association TEDAE, Spain's defense industry declined by 7.5% in 2018, reached a total turnover of €4.9Bn and employed 20,519 professionals in 388 companies. [M] Defense turnover impacted programs of Eurofighter Typhoon, the EJ2000 engine for the Eurofighter, the A400M aircraft, the NH-90 helicopter and the MTR390 engine for the TIGER combat helicopter. Despite the previous decline, Spain, in 2018, elected new government officials and now are aggressively improving their arms import and export capability. [M] Part of the New Modernization plan implied generating closer institutional relations with the purpose of moving forward that will allow for greater export as well as imports.

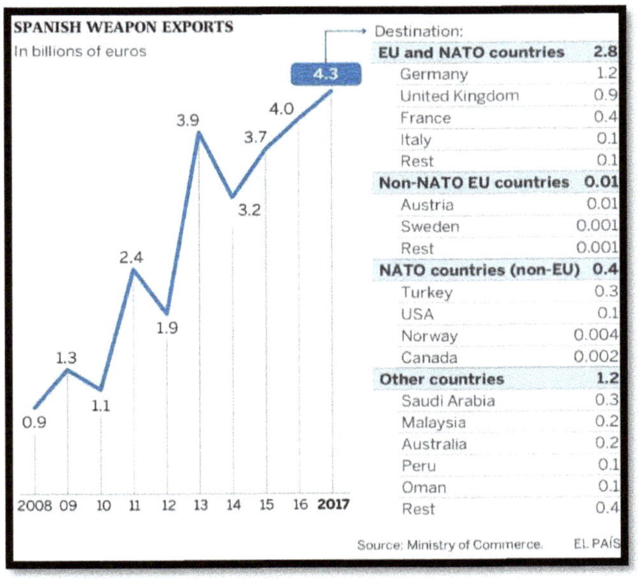

Figure 80 Spain's weapon export value and destinations.
Source: Ministry of Commerce

M The Spanish Government signed the implementation agreement for the first phase. "In a third agreement, Spain will be incorporated with a 33% stake equal to that of Germany and France", Ángel Olivares, secretary of State of Defence told ESD, "For Spain, and our European partners, it is a strategic project that will involve a technological revolution". M

To further Spain's growth into the arm trade, the Minister of Defence is Margarita Robles, an independent Socialist Party politician with a background as a judge. In the last 18 Months, the Government has approved several major projects: the five new Navantia F-110 frigates (€4.3Bn); completing the four Navantia S-80 submarines (€1.7Bn) (Figure 81) , two new military satellites HISDESAT SPAINSAT and XTAR-EUR (€1.4Bn); the procurement of 23 new Airbus NH-90 helicopters (€1.3Bn); the modernization of the Eurofighter fleet (€906M) by Airbus; and the modernization of the CHINOOK helicopters (€1Bn). M This is the rebuilding portion that will project Spain to likely increase this arms transfer by 2030. "The Spanish defense industry faces the challenge of actively participating in the consolidation of the European defense market, where further internationalization opportunities will arise. The pace at which a single European defense market is being created must be changed so that those companies that are currently less accustomed to working in this environment get used to it," says Secretary of State for Defence Ángel Olivares, highlighting the new opportunities that EU defense initiatives are likely to provide in the coming decade..

Figure 81 In 2019, the Spanish Government approved the completion of the four Navantia S-80 submarines for €1.7Bn. (Photo: Armada Espanola). Source:

Analytic Confidence:

Analytic confidence on this estimate is *moderate*. Open sources were used and were generally reliable. Source content varied; however, sources tended to corroborate one another. There was adequate time allotted to the task and the analyst used a structured research method.

Authored by: Jerry Brown

Israel's Arms Trade Likely to Increase Over the Next 10 Years Amid Global Demands for Weapons Purchases

Executive Summary:

Following decades of successful arms trades providing quality, high-grade military weapons, Israel's global arms trade is likely to moderately increase over the next 10 years. The forecasted increase is largely due to the addition of new customers including the Asian market that, combined with Europe, will boosts Israel's global arms trade. Despite the large efforts from human rights organizations to counter Israel's arms export, Israel continues to employ groundbreaking strategies to increase its arms trade that supports its domestic economy and national security interests.

Discussion:

Previous decades have shown a large increase and profitable earning on Israel arms trade. Historical data is key in forecasting future Israel arms trade earnings, and it demonstrated the highest earnings in 2017 at USD $1,441 million from 1954 to 2018. Although the actual earnings are at USD $707 million, there are still three more arms trade quarters left in 2020. [H]

Most of the world's arms-trade experts identified India, Azerbaijan, and Vietnam as the existing top Israel's arms customers. [H] It is likely that due to volatility in regional security in the Middle East region, Israel relies on a promising defense industry as a strategic source to its country's economy and national security. [M]

Figure 82 SIPRI Global Aircraft Sales

With direct arms purchases from Israel, Azerbaijan has traditionally remained one of the leading importers, and will likely continue to import arms from the Israel Defense Industry. Reports from Trading Economics totaled $46.50M in 2018 for Azerbaijan's arms purchases from Israel, with Sri Lanka listed as the lowest buyer with $2K. [H] Figure 82 from the Stockholm International Peace Research Institute (SIPRI) denotes the remarkable aircraft engineering contributing to an increase in arms trade with "three Israeli companies' arms sales of $8.7 billion accounted for 2.1 percent of the Top

100 total. Elbit Systems, Israel Aerospace Industries, and Rafael all increased their arms sales in 2018." H

Israel's arms trade increased nearly 40% in the past decade, and it is likely that new arms trades with India will increase its future arms sales. With an unprecedented $9.2 billion record arms trade sales in 2017, a major portion of Israel's profits were from India's purchases. H One of the top customers of arms trade for Israel continues to be India with diplomatic relationship stronger than ever, and even in the midst of the COVID-19, and according to a statement released by the Indian government, "Israel will supply the Indian military with 16,479 Negev light machine guns, despite growing concerns over the health emergency facing the country of 1.3 billion people." H

Other reasons for Israel's successful arms trade are the quality and the advanced technology of the military products that improve amicable diplomatic relationships between Israel and other nations. In fact, according to Kubovich, from the Haaretz Israel News, "arms deals have become an important part of Israel's relations with other countries. For example, a half-billion-dollar deal was recently signed between Israel and Croatia for the sale of F-16 aircraft." H Figure 83 from a leading Israel publication, Haaretz, depicts the major breakdown in Israel's arms trade categories by percentage. H

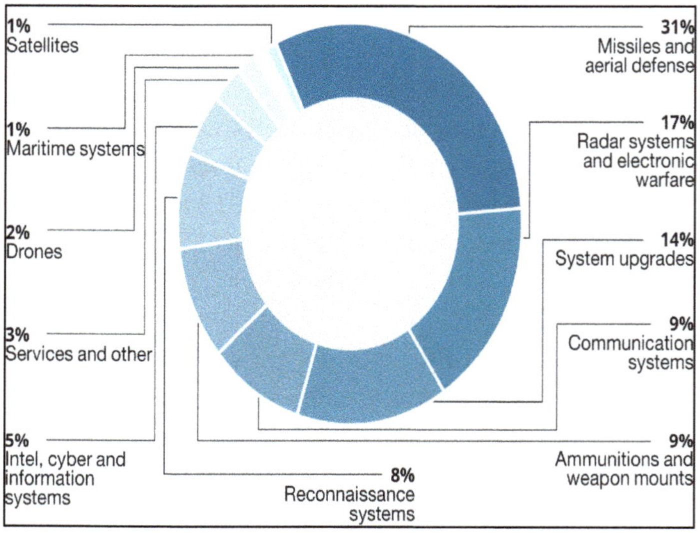

Figure 83 Haaretz Breakdown on Israel Exports 2017

The emergent Coronavirus (COVID-19) crisis may impact Israel's future arms trade sales. Although too early to identify the exact impacts of COVID-19 on future Israel arms trades, according to the Stockholm International Peace Research Institute (SIPRI) "the trade deficit in Israel decreased to USD 1,305.9 million in February of 2020 from USD 1,415.1 million in the corresponding month of the previous year. Exports dropped 6.7 percent to USD 4,390.3 million while imports fell 6.9 percent to USD 5,696.2 million." H

Opposition to Israel's arms trade continues and it is difficult to estimate the impacts on future sales. France and Germany who are rated in the top five exporters in addition to the United States (US), Russia, and China between 2014 and 2018 helped contribute to some of the largest weapons clients but do not compete against Israel's arms trade customers. H Human rights organizations constantly seek global support and published anti-arms trade rhetoric with the objective to reduce or eliminate Israel's arms exports. [H] Celebrating its 70th anniversary, the United Nations (UN) resolution 217A of 1948 provides educational material and regulations against human rights violations and encourages the immediate cease of arms sales to countries that are known for violating human rights. [H] Human rights organizations accused Israel of exporting arms sales to countries that are known to be violators of human rights. As the eighth largest arms trade in accordance to the Stockholm International Peace Research Institute (SIPRI) latest report, human rights organizations such as Amnesty International globally accused Israel of selling weapons to South Sudan, Myanmar, Mexico, and the UAE. [H]

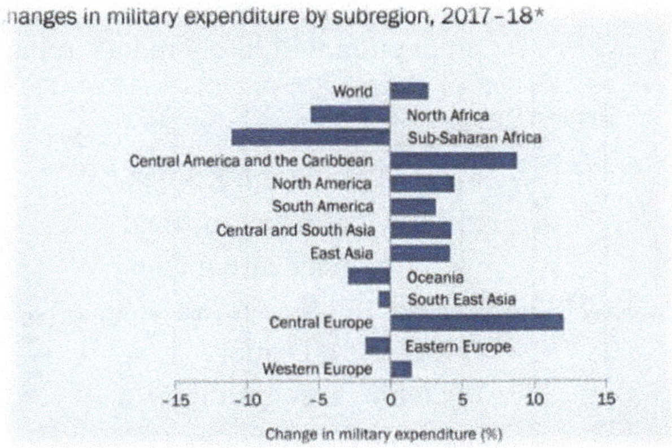

Figure 84 SIPRI 2019 Report on Global Mil Expenditures

Despite the large efforts from human rights organizations to counter Israel's arms export, Israel continues to gain remarkable profits while, according to Toi, from The Times of Israel, in March 2019 "the Defense Minister believes the increase in sales stems from several international trends, one being is an increase in defense budgets of NATO countries." [H] In addition, Israel's own legislation established parameters to ensure arms trades are not conducted with countries that endorse human rights violations and "defense exports are regulated according to a 2007 law that requires defense contractors to consider what and where the Israeli weapons will be used for." [H]

Latest reports from Haaretz supports Israel's likely increase in arms trade as the estimates hinge on how in the past "increased in its arms exports by 77 percent in 2015-2019, compared to the five previous years" [M] Lastly, other countries that currently import U.S. weapons are seeking a new trade agreement with Israel. In figure 84, a SIPRI 2019 report depicts the military expenditures changes by subregion with emerging markets that will likely contribute in increasing Israel's future arms trade. [M] A prime example is the recent and first visit from the Philippines President, Duterte, and "there could be far bigger deals on the way as Manila plans a multi-billion-dollar overhaul of its armed forces. Duterte has been dismissive of American sales overtures, saying he does not need US fighter jets or submarines." [M]

Analytic Confidence:

Analytic confidence on this estimate is *moderate* due to large variants in future economic estimates. Reliable sources provided high quality and credible data analysis. Most sources corroborated with each other including Israel's publishers. Considerable time contributed to complimenting a structured research method.

Authored by: Rafael Duran

South Korea's Arms Trade Likely to Increase Over Next 10 Years Due to Continuous North Korean Threat

Executive Summary:

Due to the growing and imminent North Korean threat, South Korea will likely increase their arms trade over the next 10 years. Despite South Korea ranking in the top nine countries for military technological expertise, it still relies on the United States (US) for most of its conventional weapons. However, by 2030 South Korea's domestic weapons production will likely reduce reliance on US weapon imports.

Discussion:

Most of the South Korean military has been outfitted with US-made armaments such as fighter aircraft (Figure 85). Daniel Pinkston, a lecturer in international relations at Troy University based in Seoul, remarked that South Korean officials expect a more even balance on future Armed Forces as South Korea increases its own weapons production because of the threat from North Korea and Washington's uncertain stewardship of stability and peace in East Asia. Pinkston also added that, in the long run, those weapons are going to be exported around the world. [M]

Figure 85: U.S.-manufactured aircraft and related weaponry accounted for 56 percent of U.S. weapons exports, according to a South Korean government agency. File Photo courtesy of U.S. Air Force

Another indicator that South Korea is moving toward a self-reliant arms trade is that the South Korean National Defense Ministry's latest budget is set to increase by seven percent yearly, meaning that the country is going to spend around $240 billion on defense from 2020 to 2024, with $85 billion going to "arms improvement." [M] While more efforts are being placed on exports, according to the Stockholm International Peace Research Institute (SIPRI), countries like Singapore, Turkey, and Spain mostly buy arms from Australia, South Korea, Kazakhstan, and Mexico. [H]

As South Korea moves towards being an independent weapon manufacturer, the country is starting to reduce import dependency for advanced military technologies. Since the 1970s, South Korea gradually has moved up the technology ladder to produce more complex systems, but they are still dependent on foreign sources for the most advanced technologies. [M] Over 60 years later, the South Korean military now possesses a formidable collection of 2,400 tanks, including the domestically produced K-1 and K-2 models, 2,300 self-propelled artillery pieces like the K-9 or K-55 and around 3,200 armored vehicles of various kinds, making it one of the most heavily armed militaries in the world M

Another indicator that South Korea will likely increase domestic arms production in 10 years comes from a study conducted by South Korea's Defense Agency for Technology and Quality (DTaQ) published by Yonhap. That study ranked South Korea as the ninth most advanced defense technology, tied with Italy, among 16 major countries as depicted in figure 86. M

Figure 86: South Korea were assessed to be the world's ninth most advanced as of last year, and about 80 percent of the U.S. levels, a state-run military research agency said Tuesday.

South Korean President Moon Jae-in's administration has committed billions of additional dollars to the defense budget, which is already among the largest in the world. In July the Ministry of National Defense announced South Korea would start build a light aircraft carrier, the country's first. And in August it unveiled a plan to spend about $239 billion more between 2020 and 2024. M About $85 billion of the future budget is earmarked for arms improvements, representing an average year-on-year increase of 10.3 percent. The planned aircraft carrier is expected to accommodate vertical-landing F-35B stealth fighter jets (Figure 87). H The same F-35B that were purchased from the US and have stealth capability. Among the other weapons on Seoul's shopping list are new missile defense systems, three more destroyers equipped with the cutting-edge Aegis radar system, spy satellites and high-altitude reconnaissance drones, anti-submarine helicopters, maritime patrol aircraft, submarines capable of firing cruise and ballistic missiles, and a warship armed with guided missiles.

A key motive for South Korea to become an independent weapons developer, producer, and exporter of military arms stems from the frustrations with the US. US defense companies do not share the latest development while demanding a huge payout from South Korea to help cover their research and development costs. [M] Also, the current US President is demanding an almost fivefold increase in South Korea's contribution to the cost of basing around 28,000 American troops in the country. [M]

Figure 87: South Korea has joined the race and announced plans to build a modified large-deck aircraft carrier based on the Republic of Korea Navy (ROKN) Dokdo Class amphibious warfare ships.

Analytic Confidence:

Analytic confidence on this estimate is *moderate*. Open sources were used and were generally reliable. Source content varied; however, sources tended to corroborate one another. There was adequate time allotted to the task and the analyst used a structured research method.

Authored by: Jerry Brown

United Kingdom's Global Arms Exports Likely to Have Modest Growth Over the Next 10 Years; Likely Increase in Military Modernization

Executive Summary:

Despite the United Kingdom's 15% drop in arms export since 2010-2014, the UK saw a record foreign sale to Saudi Arabia, UAE, and Qatar for £14B ($18M) in 2018. Due to a renewed trend towards modernization to counter a resurgent Russia and to meet NATO obligations; however, it is likely UK foreign military sales will have a modest growth trend over the next 10 years.

Discussion:

As of December 2019, arms sales are up 4.6 percent worldwide. [H] The United Kingdom (UK) ranks 6th in the global share of major arms exports (figure 88). [H] The UK's arms exports, by volume, decreased by 15% since 2010-2014 primarily as a result of decreases in its arms exports to Saudi Arabia, India and the USA (figure 89). [H] According to Stockholm International Peace Research Institute (SIPRI), the UK's top arms customers are Saudi Arabia, Oman, and the United States, the UK mostly buys arms from the US, South Korea, and Germany. [H]

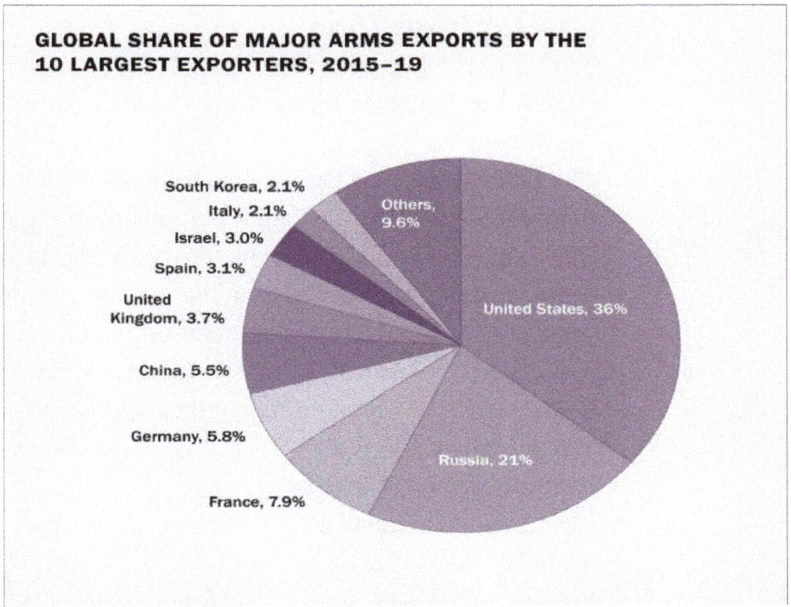

Figure 88: Global share of top arms exporters, 2015-19. Source: SIPRI

Driven by the need to counter a resurgent Russia, the UK's defense expenditures are anticipated to increase from US$50.3 billion in 2019 to US$52.6 billion in 2020. [M]

Technological advancement will upsurge the demand for the replacement of ageing equipment and the procurement of new products, will likely propel the defense market in the coming years. [M] The Conservative Party manifesto talked up defense spending, pledging to meet NATO's target of spending 2% of GDP on defense in order to keep its members safe; however, Prime Minister Boris Johnson has said he would go further than that, by increasing the defense budget by "at least 0.5% above inflation every year". [M] It is unclear how the COVID-19 pandemic may affect the proposes UK budget increases; however, it is likely that impacts will moderate planned budgetary increases. [M]

Both the UK air force and army have planned modernization efforts. The UK's air force, if funding remains as planned, is likely to experience a major growth in the UK defense market with procurements and replacement plans for combat aircraft, airborne ISR equipment, and orders for new unmanned aerial vehicles. M The UK plans to develop a next generation stealth fighter jet, the Tempest project, figure 91, and plans to accelerate in the coming 12 months. M The proposed Tempest plan gradually replaces fourth-generation Typhoon fighters and complements F-35 stealth jets. M

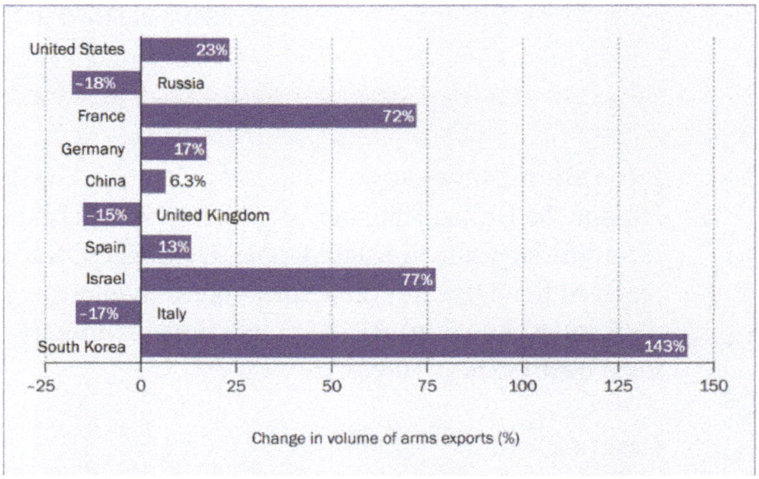

Figure 89: Change in volume of arms exports (%). Source SIPRI

The UK Ministry of Defence announced it plans to invest up to $162 million developing three directed-energy weapon demonstrators, including one aimed at killing drones. The MoD said it had notified industry this week, in what is called a Prior Information Notice, of its intention to procure two laser-based demonstrators and a radio-frequency weapon to "explore the potential of the technology and accelerate its introduction onto the battlefield." H The UK Ministry of Defense has announced additional funding for the British Army to stand up new cyber operations centers across the UK. M Speaking at the Royal United Service Institute's Land Warfare Conference, the head of the army, Mark Carleton-Smith, noted that it was "indisputably the case" that a technical revolution was underway, and that a response by the British Army needed to be equally revolutionary, "Secure borders, or living on an island, are no guarantees against the corrosive and intrusive effects of disinformation, subversion and cyber," he explained. M The army's vehicle fleet is likely to see new vehicle systems, the Ajax and Boxer, to form newly-formed Strike brigades, and upgrades to legacy systems like the Warrior Infantry Fighting

Figure 90: United Kingdom's Challenger 2 Main Battle Tank. Source: The National Interest

Vehicle (IFV) and Challenger 2 Main Battle Tank (MBT), figure 90, are ongoing.[H] Delivering a keynote speech at 'Defence IQ's International Armoured Vehicles 2020', British Army director capability Major General Jez Bennett said: "The British Army, therefore, has a modernization challenge. We are seeking to invest in new and novel technologies whilst simultaneously attempting to modernize an ageing and increasingly obsolete fleet."[H] The British Army last December confirmed it had ordered 508 Boxer Mechanized Infantry Vehicles (MIV) at a cost of £2.8bn ($3.3B) and has ordered a total of 589 Ajax platform vehicles.[H]

According to UK news sources, the government will invest at least £800m into a new "blue skies" science research agency as part of a series of pro-business measures designed to boost Britain's competitive edge.[M] The agency is part of plans to increase public investment in research and development from £11.4bn currently to £22bn per year by 2024-25 and make the UK "one of the scientific and research centres of the world".[M]

Figure 91: A prototype mockup of the Tempest stealth fighter. Source: The National Interest

It is likely that the UK will see a modest increase in arms export trends as the UK seeks to revitalize the defense industry sectors and improve productivity. The UK launched a review, led by the Ministry of Defence (MoD), to be complete in 2020, perhaps later due

Export credit business supported (medium/long-term) 2018-19

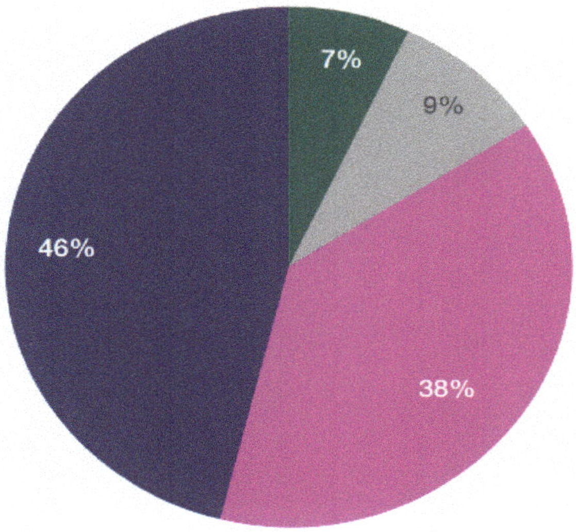

Sector	UKEF maximum liability, £m
Aerospace	482
Civil - direct loans	587
Civil - guarantees	2,542
Defence	3,042

Figure 92: Business sectors supported by UKEF financing. Source: UK Export Finance

to COVID-19,[M] will identify how the government can take a more strategic approach to ensure that the UK has competitive, innovative and world-class defence and security industry sectors. [M] UK Defence Secretary Ben Wallace said, "Our relationship with industry is crucial to maintaining the UK's position as a Tier 1 military power." [M] UK Defence Minister Jeremy Quin add a strategic goal of the report as, "On the international stage, UK defence and security companies play a crucial role in maintaining the UK's global influence, underpinning our strategic partnerships with key allies." [M] Helping to fulfill strategic partnerships*, the* UK Export Finance (UKEF) is the UK's export credit agency and a government department, strategically and operationally aligned with the Department for International Trade. UKEF underwrites guarantees and loans, where the private sector will not, using UK government-backed funds, figure 92.[H] It is likely the UKEF will continue strengthening the UK industrial base and securing strategic ally support. If future trends continue, it is likely most export sales will be with aircraft, vessels and missiles. Support from UKEF will help sustain the UK's defence industrial capability into the future. [H]

Analytic Confidence:

Analytic confidence on this estimate is *moderate*. Open sources were used and were generally reliable. Source content varied; however, sources tended to corroborate one another. Volatility in pandemic impacts may affect the estimate. There was adequate time allotted to the task and the analyst used a structured research method.

Authored by: Russell Hoff

ESTIMATIVE CHARTS

GLOBAL ESTIMATED AVERAGE THREAT MODERNIZATION AND FUTURE CAPABILITY MATRIX

Certainty 100%		
Almost Certain	86-99%	Likelihood
Highly Likely	71-85%	
Likely	56-70%	
Chances a Little Better [or Less]	46-55%	
Unlikely	31-45%	
Highly Unlikely	16-30%	
Remote	1-15%	
Impossibility 0%		

findings on estimated threats to the U.S. Army in the next 10 to 20 years. The most likely threats posed will be Anti-satellite operations, Cyber, Electronic Warfare, Hypersonic's, Quantum Computing, Short-range strike capability, and unmanned systems. The x-Axis (Vertical) represents a "0" to "8" numerical system where 0 = Impossible, 1 = Remote, 2 = Highly Likely, 3 = Unlikely, 4 = Chances a Little Better [or Less], 5 = Likely, 6 = Highly Likely, 7 = Almost Certain, and 8 = Certain.

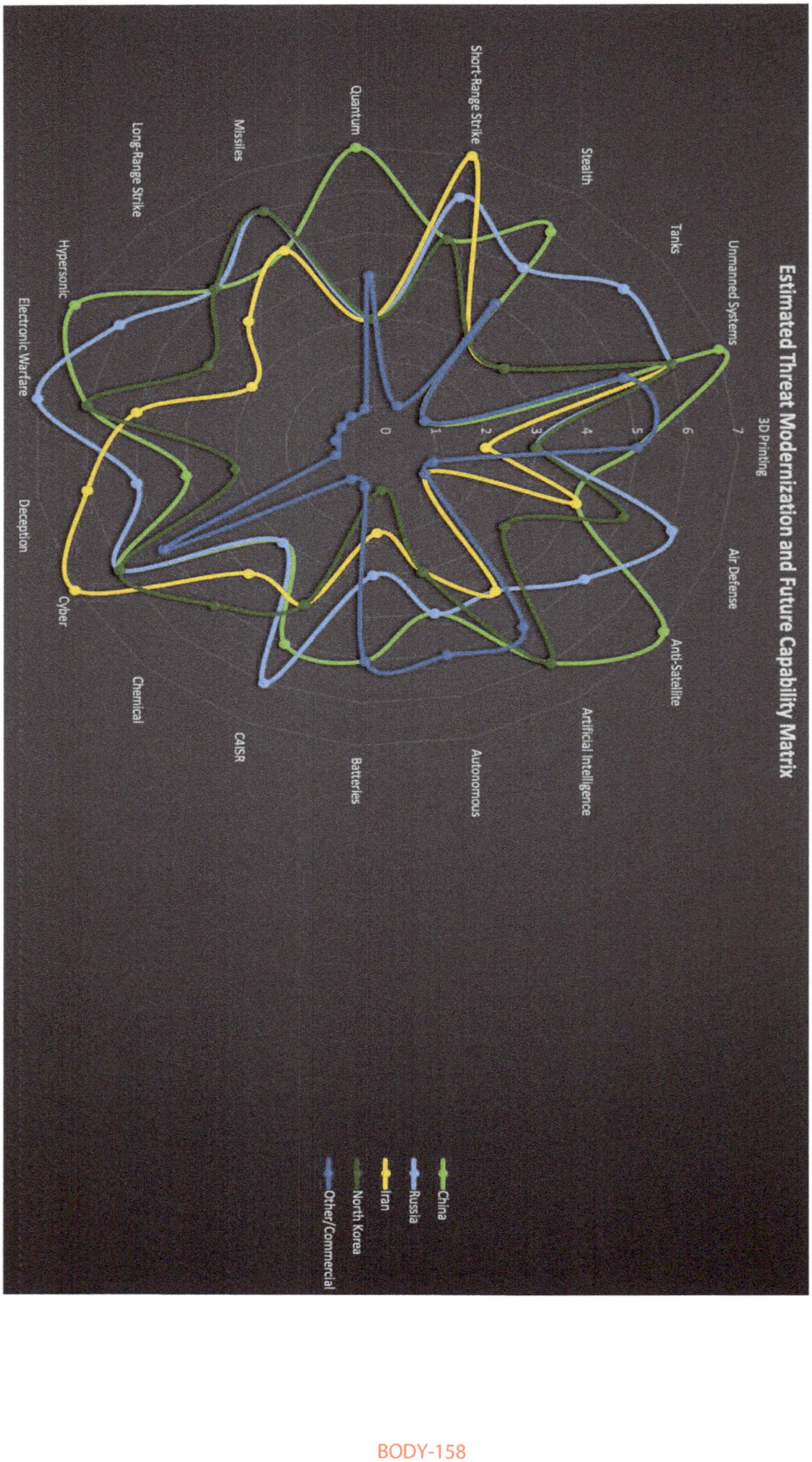

CHINA'S ESTIMATED FUTURE SIGNATURE AND THREAT CAPABILITY MATRIX (prior to aggregate cutoff)

The below chart leverages Team Sensing's Difficulty Scale which provides their analytical perspective of US Difficulty to detect the signatures on the platforms in 10 to 20 years. The x-Axis (Vertical) represents a "0" to "8" numerical system where 0 = Impossible, 1 = Remote, 2 = Highly Likely, 3 = Unlikely, 4 = Chances a Little Better [or Less], 5 = Likely, 6 = Highly Likely, 7 = Almost Certain, and 8 = Certain.

Estimated Signature and Capability Threat Matrix - China

	C4ISR	Electronic Warfare	Hypersonic	Stealth	Autonomous	Anti-Satellite	Unmanned Systems	Artificial Intelligence	Cyber	Quantum	Deception	Air Defense	Long-Range Strike	Missiles	Short-Range Strike	3D Printing	Batteries	Chemical	Tanks
Electronic (Digital/Software)	4	4	3	3	3	3	3	4	4	4	3	4	3	3	3	3	3	3	3
Frequency	4	4	4	4	4	3	4	3	3	3	3	3	3	3	3	3	3	3	3
Radio waves	4	4	4	4	4	3	4	3	3	3	3	3	3	3	3	3	3	3	3
Doctrine & Non-Technical Process	4	2	4	4	3	4	3	3	3	4	4	4	3	3	3	3	3	3	3
Digital	4	4	3	3	3	3	3	4	4	4	3	3	3	3	3	3	3	3	3
Physical	4	4	4	3	3	4	4	3	3	3	4	2	3	3	3	3	3	3	3
Visual	4	4	3	3	4	3	4	3	3	3	4	2	3	3	3	3	3	3	3
Wavelength	4	4	3	4	4	3	3	3	3	3	3	3	3	3	3	3	3	3	3
Radiation	4	4	3	3	3	3	3	3	3	3	3	4	3	3	3	3	3	3	3
Metanalytic	3	4	3	3	3	3	3	4	4	3	3	3	3	3	3	3	3	3	3
Speed	4	3	3	3	4	3	3	3	3	3	3	3	3	3	3	3	3	3	3
Hyperspectral	4	3	3	3	3	3	3	3	3	3	3	4	3	3	3	3	3	3	3
Thermal	2	2	4	4	3	4	3	3	3	3	3	3	3	3	3	3	3	3	3
Laser	3	3	3	3	3	4	3	3	3	3	3	3	3	3	3	3	3	3	3
Infrared	2	2	4	4	3	4	3	3	3	3	3	3	3	3	3	3	3	3	3
Acoustic	3	3	3	3	3	3	3	3	3	3	3	3	3	3	3	3	3	3	3
Bio/Chemical	3	3	3	3	3	3	3	3	3	3	3	3	3	3	3	3	3	3	3
Magnetic	3	3	3	3	3	3	3	3	3	3	3	3	3	3	3	3	3	3	3
Seismic	3	3	3	3	3	3	3	3	3	3	3	3	3	3	3	3	3	3	3
Smell	3	3	3	3	3	3	3	3	3	3	3	3	3	3	3	3	3	3	3
Ultra Violet	3	3	3	3	3	3	3	3	3	3	3	3	3	3	3	3	3	3	3
X-Ray	3	3	3	3	3	3	3	3	3	3	3	3	3	3	3	3	3	3	3

5 Much More Difficult to Detect	4 More Difficult to Detect	3 Moderate to Routine Improvement	2 Easier to Detect	1 Extremely Easier to Detect

ESTIMATED FUTURE SIGNATURE AND THREAT CAPABILITY MATRIX – CHINA (after aggregate cutoff)

The below chart leverages Team Sensing's Difficulty Scale which provides their analytical perspective of US Difficulty to detect the signatures on the platforms in 10 to 20 years. The x-Axis (Vertical) represents a "0" to "8" numerical system where 0 = Impossible, 1 = Remote, 2 = Highly Likely, 3 = Unlikely, 4 = Chances a Little Better [or Less], 5 = Likely, 6 = Highly Likely, 7 = Almost Certain, and 8 = Certain.

Estimated Signature and Capability Threat Matrix - China

	C4ISR	Electronic Warfare	Hypersonic	Stealth	Autonomous	Anti-Satellite	Unmanned Systems	Artificial Intelligence	Cyber	Quantum	Deception
Frequency	4	4	4	4	4	3	4	3	3	3	3
Radio waves	4	4	4	4	4	3	4	3	3	3	3
Physical	4	4	4	3	3	4	4	3	3	3	4
Electronic (Digital/Software)	4	4	3	3	3	3	3	4	4	4	3
Doctrine & Other Non-Technical Process	4	2	4	4	3	4	3	3	3	4	4
Digital	4	4	3	3	3	3	3	4	4	4	3
Visual	4	4	3	3	4	3	4	3	3	3	4
Wavelength	4	4	3	4	4	3	3	3	3	3	3
Metanalytic	3	4	3	3	3	3	3	4	4	3	3
Radiation	4	4	3	3	3	3	3	3	3	3	3
Speed	4	3	3	3	4	3	3	3	3	3	3
Hyperspectral	4	3	3	3	3	3	3	3	3	3	3
Thermal	2	2	4	4	3	4	3	3	3	3	3
Laser	3	3	3	3	3	4	3	3	3	3	3
Infrared	2	2	4	4	3	4	3	3	3	3	3
Acoustic	3	3	3	3	3	3	3	3	3	3	3
Bio/Chemical	3	3	3	3	3	3	3	3	3	3	3
Magnetic	3	3	3	3	3	3	3	3	3	3	3
Seismic	3	3	3	3	3	3	3	3	3	3	3
Smell	3	3	3	3	3	3	3	3	3	3	3
Ultra Violet	3	3	3	3	3	3	3	3	3	3	3
X-Ray	3	3	3	3	3	3	3	3	3	3	3

5 Much More Difficult to Detect	4 More Difficult to Detect	3 Moderate to Routine Improvement	2 Easier to Detect	1 Extremely Easier to Detect

ESTIMATED FUTURE SIGNATURE AND THREAT CAPABILITY MATRIX – RUSSIA (prior to aggregate cutoff)

The below chart leverages Team Sensing's Difficulty Scale which provides their analytical perspective of US Difficulty to detect the signatures on the platforms in 10 to 20 years. The x-Axis (Vertical) represents a "0" to "8" numerical system where 0 = Impossible, 1 = Remote, 2 = Highly Likely, 3 = Unlikely, 4 = Chances a Little Better [or Less], 5 = Likely, 6 = Highly Likely, 7 = Almost Certain, and 8 = Certain.

Estimated Signature and Capability Threat Matrix - Russia

	Electronic Warfare	Deception	Cyber	C4ISR	Hypersonic	Air Defense	Missiles	Tanks	Stealth	Autonomous	Artificial Intelligence	Unmanned Systems	Batteries	Anti-Satellite	Long-Range Strike	Quantum	Chemical	3D Printing	Short-Range Strike
Electronic (Digital/Software)	4	4	4	4	3	4	4	4	3	3	4	3	3	3	3	3	3	3	3
Visual	4	4	4	4	4	2	3	5	5	4	3	3	3	3	3	3	3	3	2
Non-Technical (Process/Doctrine)	5	5	4	4	4	4	2	3	3	3	3	4	3	3	3	3	3	3	3
Digital	4	4	4	4	3	3	4	3	3	3	4	4	3	3	3	3	3	3	3
Radio waves	4	4	4	4	4	4	4	3	3	3	3	3	3	3	3	3	3	3	3
Speed	4	4	4	4	4	3	3	4	4	4	3	2	3	3	3	3	3	3	3
Magnetic	4	4	4	4	3	3	4	4	3	3	3	3	3	3	3	3	3	3	3
Physical	4	4	4	4	3	4	3	3	3	4	3	3	3	3	3	3	3	3	3
Frequency	4	4	4	4	4	3	4	3	3	3	3	4	3	3	3	3	3	3	2
Thermal	4	4	4	4	4	4	2	4	4	3	3	3	3	3	3	3	3	3	2
Acoustic	4	4	3	3	3	3	4	3	4	3	3	3	3	3	3	3	3	3	3
Hyperspectral	4	4	4	4	3	2	4	3	3	3	3	3	3	3	3	3	3	3	3
Radiation	4	3	3	3	3	4	4	3	3	3	3	3	3	3	3	3	3	3	3
Wavelength	4	4	4	4	3	3	3	3	3	3	3	3	3	3	3	3	3	3	2
x-Ray	4	3	3	3	3	3	3	3	3	3	3	3	3	3	3	3	3	3	3
Laser	3	3	3	3	3	3	3	3	3	3	3	3	3	3	3	3	3	3	3
Smell	3	3	3	3	3	3	3	3	3	3	3	3	3	3	3	3	3	3	3
Metanalytic	3	3	3	3	3	3	3	3	3	3	3	3	3	3	3	3	3	3	3
Ultra Violet	3	3	3	3	3	3	3	3	3	3	3	3	3	3	3	3	3	3	3
Seismic	3	3	3	3	4	3	3	2	3	3	3	3	3	3	3	3	3	3	3
Infrared	4	2	2	2	3	4	2	4	3	3	3	3	3	3	3	3	3	3	3
Bio/Chemical	3	3	3	3	3	3	3	3	3	3	3	3	3	3	3	3	3	2	3

5 Much More Difficult to Detect	4 More Difficult to Detect	3 Moderate to Routine Improvement	2 Easier to Detect	1 Extremely Easier to Detect

ESTIMATED FUTURE SIGNATURE AND THREAT CAPABILITY MATRIX – RUSSIA (after aggregate cutoff)

The below chart leverages Team Sensing's Difficulty Scale which provides their analytical perspective of US Difficulty to detect the signatures on the platforms in 10 to 20 years. The x-Axis (Vertical) represents a "0" to "8" numerical system where 0 = Impossible, 1 = Remote, 2 = Highly Likely, 3 = Unlikely, 4 = Chances a Little Better [or Less], 5 = Likely, 6 = Highly Likely, 7 = Almost Certain, and 8 = Certain.

Estimated Signature and Capability Threat Matrix - Russia

	Air Defense	C4ISR	Cyber	Deception	Electronic Warfare	Hypersonic	Missiles	Stealth	Tanks
Acoustic	3	3	3	4	4	3	4	4	3
Bio/Chemical	3	3	3	3	3	3	3	3	3
Digital	3	4	4	4	4	3	4	3	3
Electronic (Digital/Software)	4	4	4	4	4	3	4	3	4
Frequency	3	4	4	4	4	4	4	3	3
Hyperspectral	2	4	4	4	4	3	4	3	3
Infrared	4	2	2	2	4	3	2	3	4
Laser	3	3	3	3	3	3	3	3	3
Magnetic	3	4	4	4	4	3	4	3	4
Non-Technical (Process/Doctrine)	4	4	4	5	5	4	2	3	3
Physical	4	4	4	4	4	3	3	3	3
Radiation	4	3	3	3	4	3	4	3	3
Radio waves	4	4	4	4	4	4	4	3	3
Seismic	3	3	3	3	3	4	3	3	2
Smell	3	3	3	3	3	3	3	3	3
Metanalytic	3	3	3	3	3	3	3	3	3
Speed	3	4	4	4	4	4	3	4	4
Thermal	4	4	4	4	4	4	2	4	4
Ultra Violet	3	3	3	3	3	3	3	3	3
Visual	2	4	4	4	4	4	3	5	5
Wavelength	3	4	4	4	4	3	3	3	3
x-Ray	3	3	3	3	4	3	3	3	3

5 Much More Difficult to Detect | 4 More Difficult to Detect | 3 Moderate to Routine Improvement | 2 Easier to Detect | 1 Extremely Easier to Detect

ESTIMATED FUTURE SIGNATURE AND THREAT CAPABILITY MATRIX – RUSSIA (after aggregate cutoff x2)

The below chart leverages Team Sensing's Difficulty Scale which provides their analytical perspective of US Difficulty to detect the signatures on the platforms in 10 to 20 years. The x-Axis (Vertical) represents a "0" to "8" numerical system where 0 = Impossible, 1 = Remote, 2 = Highly Likely, 3 = Unlikely, 4 = Chances a Little Better [or Less], 5 = Likely, 6 = Highly Likely, 7 = Almost Certain, and 8 = Certain.

Estimated Signature Difficult to Detect No Noise - Russia

	C4ISR	Cyber	Deception	Electronic Warfare
Digital	4	4	4	4
Electronic (Digital/Software)	4	4	4	4
Frequency	4	4	4	4
Hyperspectral	4	4	4	4
Magnetic	4	4	4	4
Non-Technical (Process/Doctrine)	4	4	5	5
Physical	4	4	4	4
Radio waves	4	4	4	4
Speed	4	4	4	4
Thermal	4	4	4	4
Visual	4	4	4	4
Wavelength	4	4	4	4

5 Much More Difficult to Detect	4 More Difficult to Detect	3 Moderate to Routine Improvement	2 Easier to Detect	1 Extremely Easier to Detect

ESTIMATED FUTURE SIGNATURE AND THREAT CAPABILITY MATRIX – IRAN (prior to aggregate cutoff)

The below chart leverages Team Sensing's Difficulty Scale which provides their analytical perspective of US Difficulty to detect the signatures on the platforms in 10 to 20 years. The x-Axis (Vertical) represents a "0" to "8" numerical system where 0 = Impossible, 1 = Remote, 2 = Highly Likely, 3 = Unlikely, 4 = Chances a Little Better [or Less], 5 = Likely, 6 = Highly Likely, 7 = Almost Certain, and 8 = Certain.

Estimated Signature and Capability Threat Matrix - Iran

	A	B	C	D	E	F	G	H	I	J	K	L	M	N	O	P	Q	R	S	T
2		Cyber	Unmanned Systems	Electronic Warfare	Deception	Air Defense	Chemical	Short-Range Strike	Artificial Intelligence	Batteries	Quantum	3D Printing	Anti-Satellite	Long-Range Strike	Missiles	Autonomous	C4ISR	Hypersonic	Stealth	Tanks
3	Doctrine and other Non-Technical Process	4	3	3	4	3	3	3	3	3	3	3	3	3	3	3	4	2	2	2
4	Physical	4	3	3	3	3	3	4	3	3	3	3	3	3	3	3	3	2	2	2
5	Acoustic	3	4	4	4	3	3	2	3	3	3	3	3	3	3	3	2	2	2	2
6	Frequency	4	4	4	3	3	3	4	3	3	3	3	3	3	3	2	1	2	2	2
7	Radiation	3	4	3	3	3	3	3	3	3	3	3	3	3	3	3	3	2	2	2
8	Speed	4	2	3	4	3	3	3	3	3	3	3	3	3	3	3	3	2	2	2
9	Thermal	4	3	3	3	3	3	4	3	3	3	3	3	3	3	3	2	2	2	2
10	Digital	4	4	3	4	3	3	3	3	3	3	2	3	2	2	3	3	2	2	2
11	Electronic (Digital/Software)	4	4	3	3	3	3	4	3	3	3	3	3	3	3	2	1	2	2	2
12	Hyperspectral	4	3	3	3	3	3	3	3	3	3	3	3	3	3	2	3	2	2	2
13	Laser	3	3	3	3	3	3	3	3	3	3	3	3	3	3	3	3	2	2	2
14	Magnetic	3	3	4	3	3	3	2	3	3	3	3	3	3	3	3	3	2	2	2
15	Metanalytic	3	3	3	3	3	3	3	3	3	3	3	3	3	3	3	3	2	2	2
16	Smell	3	3	3	3	3	3	3	3	3	3	3	3	3	3	3	3	2	2	2
17	Ultra Violet	3	3	3	3	3	3	3	3	3	3	3	3	3	3	3	3	2	2	2
18	Visual	3	3	3	4	3	3	2	3	3	3	3	3	3	3	3	3	2	2	2
19	Wavelength	3	3	4	3	3	3	2	3	3	3	3	3	3	3	3	3	2	2	2
20	x-Ray	3	3	3	3	3	3	3	3	3	3	3	3	3	3	3	3	2	2	2
21	Seismic	3	3	3	3	3	3	3	3	3	3	3	3	3	3	3	2	2	2	2
22	Infrared	2	4	3	2	3	3	3	3	3	3	3	3	3	3	2	2	2	2	2
23	Radio waves	3	3	3	2	3	3	3	3	3	3	3	3	3	3	2	2	2	2	2
24	Bio/Chemical	2	4	2	2	3	3	3	2	2	2	2	1	2	2	2	2	2	2	2

5 Much More Difficult to Detect	4 More Difficult to Detect	3 Moderate to Routine Improvement	2 Easier to Detect	1 Extremely Easier to Detect

ESTIMATED FUTURE SIGNATURE AND THREAT CAPABILITY MATRIX – IRAN (after aggregate cutoff)

The below chart leverages Team Sensing's Difficulty Scale which provides their analytical perspective of US Difficulty to detect the signatures on the platforms in 10 to 20 years. The x-Axis (Vertical) represents a "0" to "8" numerical system where 0 = Impossible, 1 = Remote, 2 = Highly Likely, 3 = Unlikely, 4 = Chances a Little Better [or Less], 5 = Likely, 6 = Highly Likely, 7 = Almost Certain, and 8 = Certain.

Estimated Signature and Capability Threat Matrix - Iran

	Cyber	Unmanned Systems	Electronic Warfare	Deception	Short-Range Strike	Air Defense	Chemical	Artificial Intelligence	Batteries	Quantum	3D Printing	Anti-Satellite	Long-Range Strike	Missiles
Frequency	4	4	4	3	4	3	3	3	3	3	3	3	3	3
Electronic (Digital/Software)	4	4	3	3	4	3	3	3	3	3	3	3	3	3
Acoustic	3	4	4	4	2	3	3	3	3	3	3	3	3	3
Doctrine and other Non-Technical Process	4	3	3	4	3	3	3	3	3	3	3	3	3	3
Physical	4	3	3	3	4	3	3	3	3	3	3	3	3	3
Thermal	4	3	3	3	4	3	3	3	3	3	3	3	3	3
Hyperspectral	4	3	3	3	3	3	3	3	3	3	3	3	3	3
Radiation	3	4	3	3	3	3	3	3	3	3	3	3	3	3
Speed	4	2	3	4	3	3	3	3	3	3	3	3	3	3
Laser	3	3	3	3	3	3	3	3	3	3	3	3	3	3
Magnetic	3	3	4	3	2	3	3	3	3	3	3	3	3	3
Metanalytic	3	3	3	3	3	3	3	3	3	3	3	3	3	3
Seismic	3	3	3	3	3	3	3	3	3	3	3	3	3	3
Smell	3	3	3	3	3	3	3	3	3	3	3	3	3	3
Ultra Violet	3	3	3	3	3	3	3	3	3	3	3	3	3	3
Visual	3	3	3	4	2	3	3	3	3	3	3	3	3	3
Wavelength	3	3	4	3	2	3	3	3	3	3	3	3	3	3
x-Ray	3	3	3	3	3	3	3	3	3	3	3	3	3	3
Digital	4	4	3	4	3	3	3	3	3	3	2	3	2	2
Infrared	2	4	3	2	3	3	3	3	3	3	3	3	3	3
Radio waves	3	3	3	2	3	3	3	3	3	3	3	3	3	3
Bio/Chemical	2	4	2	2	3	3	3	2	2	2	2	1	2	2

5 Much More Difficult to Detect	4 More Difficult to Detect	3 Moderate to Routine Improvement	2 Easier to Detect	1 Extremely Easier to Detect

ESTIMATED FUTURE SIGNATURE AND THREAT CAPABILITY MATRIX – N. KOREA (prior to aggregate cutoff)

The below chart leverages Team Sensing's Difficulty Scale which provides their analytical perspective of US Difficulty to detect the signatures on the platforms in 10 to 20 years. The x-Axis (Vertical) represents a "0" to "8" numerical system where 0 = Impossible, 1 = Remote, 2 = Highly Likely, 3 = Unlikely, 4 = Chances a Little Better [or Less], 5 = Likely, 6 = Highly Likely, 7 = Almost Certain, and 8 = Certain.

Estimated Signature and Capability Threat Matrix - North Korea

	Missiles	Long-Range Strike	Short-Range Strike	Stealth	Artificial Intelligence	Electronic Warfare	Unmanned Systems	Cyber	3D Printing	Air Defense	Anti-Satellite	Autonomous	Batteries	C4ISR	Chemical	Deception	Hypersonic	Quantum	Tanks
Physical	4	4	4	2	3	3	3	3	2	2	2	2	2	2	2	2	2	2	2
Visual	4	4	4	2	3	3	3	3	2	2	2	2	2	2	2	2	2	2	2
Speed	4	4	4	2	3	3	3	3	2	2	2	2	2	2	2	2	2	2	2
Electronic (Digital/Software)	3	3	3	2	3	3	3	3	2	2	2	2	2	2	2	2	2	2	2
Radio waves	3	3	3	2	3	3	3	3	2	2	2	2	2	2	2	2	2	2	2
Frequency	3	2	2	2	3	3	3	3	2	2	2	2	2	2	2	2	2	2	2
Acoustic	3	3	3	2	3	2	2	3	2	2	2	2	2	2	2	2	2	2	2
Thermal	4	3	3	2	2	2	2	3	2	2	2	2	2	2	2	2	2	2	2
Digital	3	3	3	2	2	2	2	3	2	2	2	2	2	2	2	2	2	2	2
Wavelength	2	2	2	2	3	3	3	3	2	2	2	2	2	2	2	2	2	2	2
Bio/Chemical	3	3	3	2	2	2	2	3	2	2	2	2	2	2	2	2	2	2	2
Infrared	3	3	3	2	2	2	2	3	2	2	2	2	2	2	2	2	2	2	2
Laser	3	3	3	2	2	2	2	3	2	2	2	2	2	2	2	2	2	2	2
Doctrine and other Non-Technical Process	3	3	3	2	2	2	2	3	2	2	2	2	2	2	2	2	2	2	2
Radiation	3	3	3	2	2	2	2	3	2	2	2	2	2	2	2	2	2	2	2
Seismic	3	3	3	2	2	2	2	3	2	2	2	2	2	2	2	2	2	2	2
Smell	3	3	3	2	2	2	2	3	2	2	2	2	2	2	2	2	2	2	2
Metanalytic	3	3	3	2	2	2	2	3	2	2	2	2	2	2	2	2	2	2	2
Ultra Violet	3	3	3	2	2	2	2	3	2	2	2	2	2	2	2	2	2	2	2
Hyperspectral	2	3	3	2	2	2	2	3	2	2	2	2	2	2	2	2	2	2	2
Magnetic	2	3	3	2	2	2	2	3	2	2	2	2	2	2	2	2	2	2	2
x-Ray	2	3	3	2	2	2	2	3	2	2	2	2	2	2	2	2	2	2	2

5 Much More Difficult to Detect | 4 More Difficult to Detect | 3 Moderate to Routine Improvement | 2 Easier to Detect | 1 Extremely Easier to Detect

ESTIMATED FUTURE SIGNATURE AND THREAT CAPABILITY MATRIX – N. KOREA (after aggregate cutoff)

The below chart leverages Team Sensing's Difficulty Scale which provides their analytical perspective of US Difficulty to detect the signatures on the platforms in 10 to 20 years. The x-Axis (Vertical) represents a "0" to "8" numerical system where 0 = Impossible, 1 = Remote, 2 = Highly Likely, 3 = Unlikely, 4 = Chances a Little Better [or Less], 5 = Likely, 6 = Highly Likely, 7 = Almost Certain, and 8 = Certain.

Estimated Signature and Capability Threat Matrix - North Korea

	Short-Range Strike	Missiles	Long-Range Strike	Artificial Intelligence	Electronic Warfare	Unmanned Systems	Cyber
Physical	4	4	4	3	3	3	3
Visual	4	4	4	3	3	3	3
Speed	4	4	4	3	3	3	3
Electronic (Digital/Software)	3	3	3	3	3	3	3
Radio waves	3	3	3	3	3	3	3
Frequency	2	3	2	3	3	3	3
Digital	3	3	3	2	2	2	3
Thermal	3	4	3	2	2	2	3
Wavelength	3	2	2	3	3	3	3
Acoustic	3	3	3	2	2	2	3
Bio/Chemical	3	3	3	2	2	2	3
Infrared	3	3	3	2	2	2	3
Laser	3	3	3	2	2	2	3
Doctrine and other Non-Technical Process	3	3	3	2	2	2	3
Radiation	3	3	3	2	2	2	3
Seismic	3	3	3	2	2	2	3
Smell	3	3	3	2	2	2	3
Metanalytic	3	3	3	2	2	2	3
Ultra Violet	3	3	3	2	2	2	3
Hyperspectral	3	2	3	2	2	2	3
Magnetic	3	2	3	2	2	2	3
x-Ray	3	2	3	2	2	2	3

5 Much More Difficult to Detect | 4 More Difficult to Detect | 3 Moderate to Routine Improvement | 2 Easier to Detect | 1 Extremely Easier to Detect

ESTIMATED THREAT MODERNIZATION AND FUTURE CAPABILITY MATRIX PER COUNTRY (CHINA and RUSSIA)

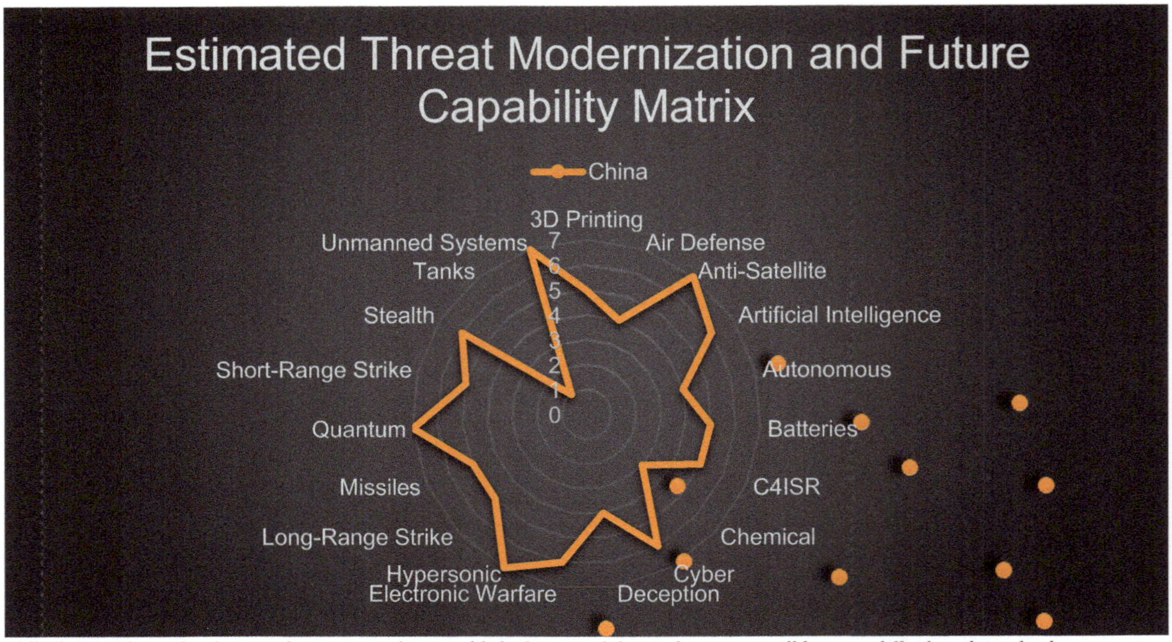

Overall Chinese Threat the closer to the outer rim the more likely their capability and signature will be more difficult to detect by the US in 10 to 20 Years

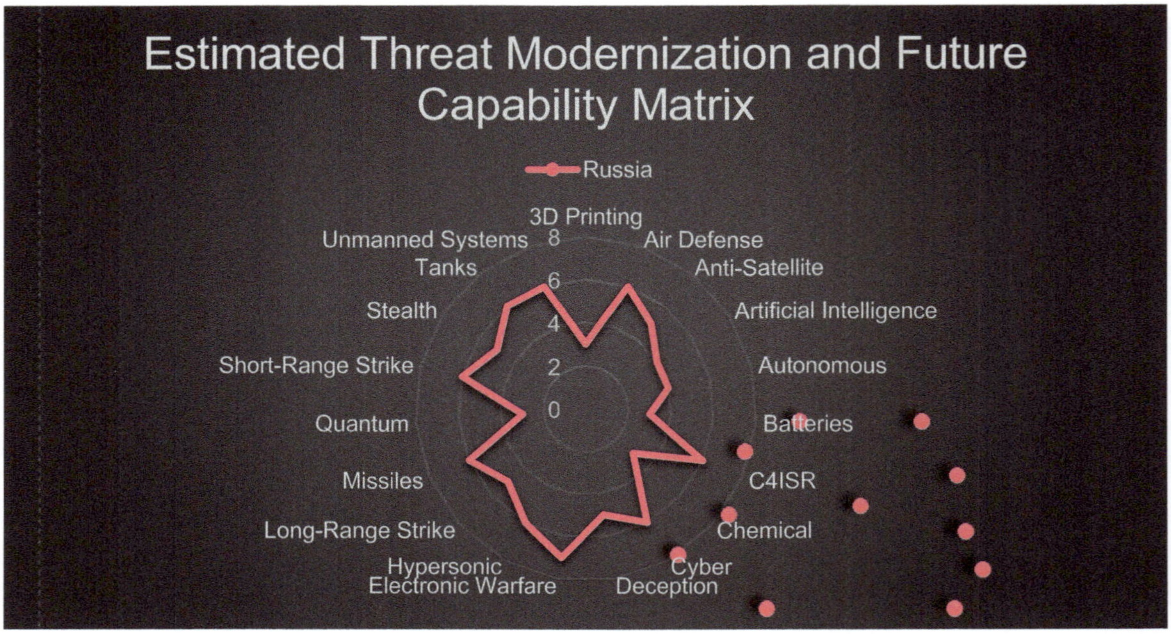

Overall Russian Threat the closer to the outer rim the more likely their capability and signature will be more difficult to detect by the US in 10 to 20 Years

ESTIMATED THREAT MODERNIZATION AND FUTURE CAPABILITY MATRIX PER COUNTRY (IRAN and NORTH KOREA)

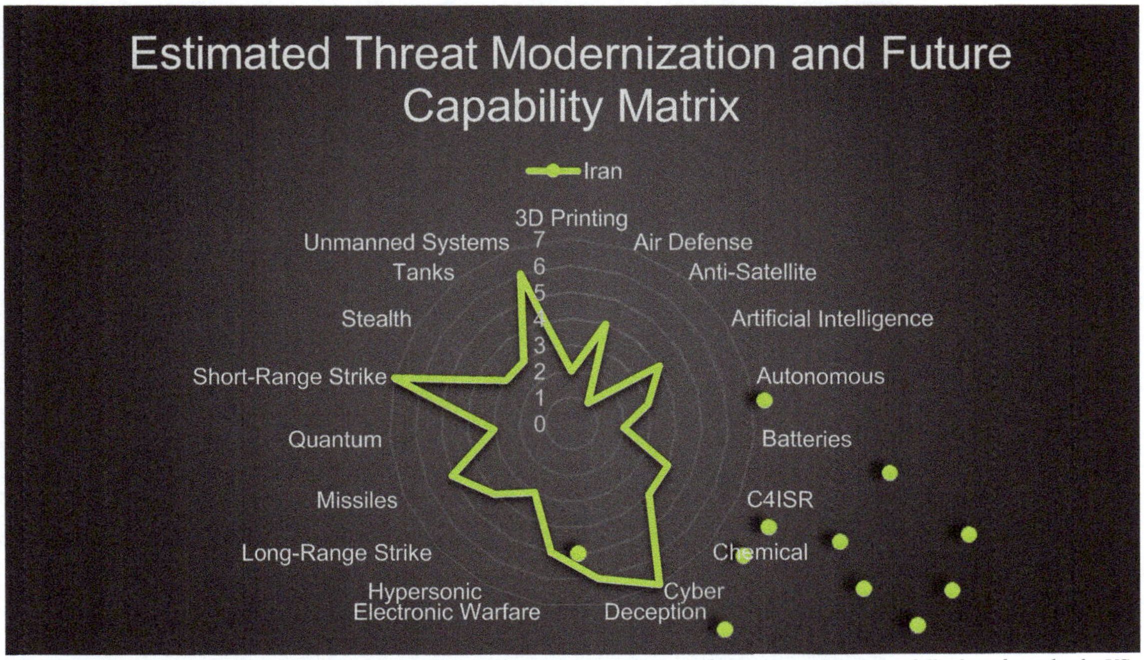

Overall Iranian Threat the closer to the outer rim the more likely their capability and signature will be more difficult to detect by the US in 10 to 20 Years

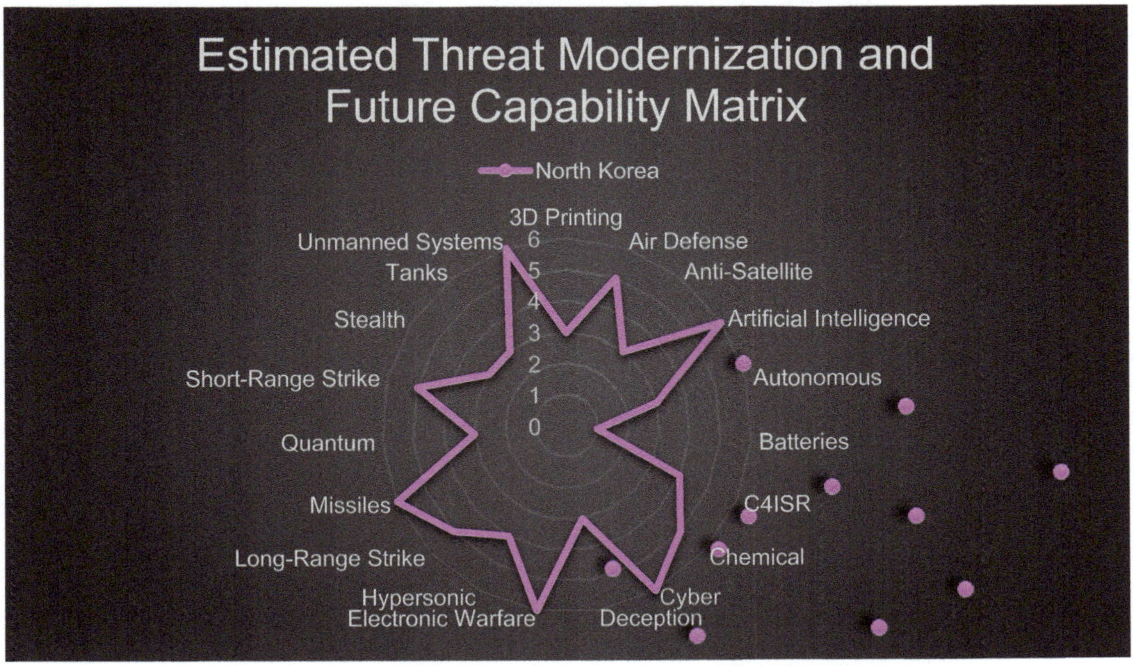

Overall North Korean Threat the closer to the outer rim the more likely their capability and signature will be more difficult to detect by the US in 10 to 20 Years

ESTIMATED THREAT MODERNIZATION AND FUTURE CAPABILITY MATRIX PER COUNTRY (OTHER and OVERALL)

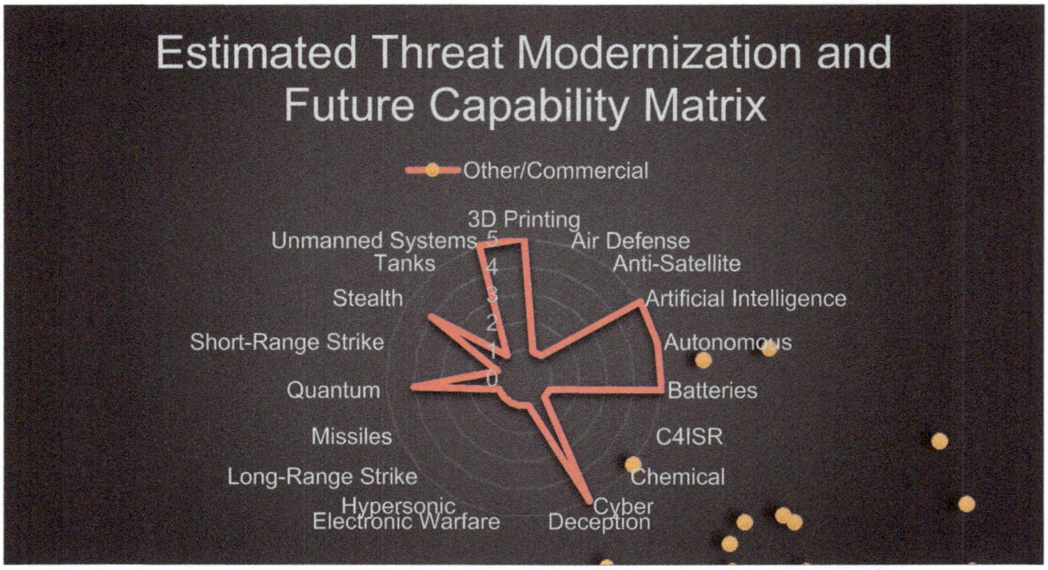

Overall Commercial and Other Country Threat's the closer to the outer rim the more likely their capability and signature will be more difficult to detect by the US in 10 to 20 Years

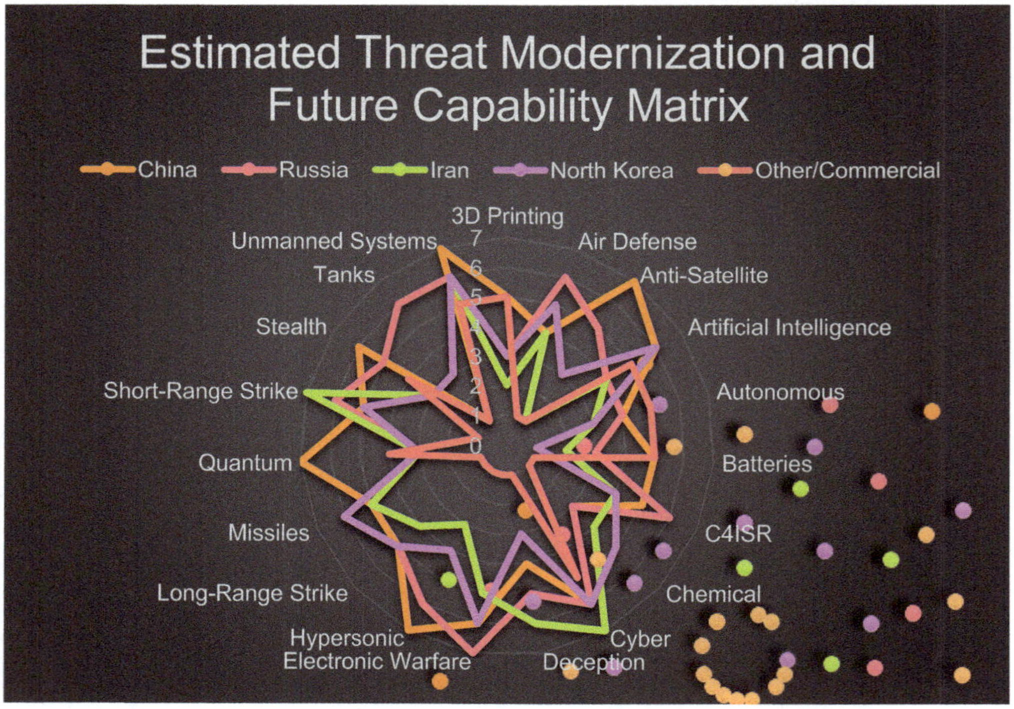

Overall World Threat - the closer to the outer rim the more likely their capability and signature will be more difficult to detect by the US in 10 to 20 Years

ESTIMATED TECHNOLOGY MODERNIZATION AND FUTURE CAPABILITY COMPARISON PER PLATFORM

Outer rim represents higher likeliness of capability and signature being more difficult to detect by the US in 10 to 20 Years

3D PRINTING

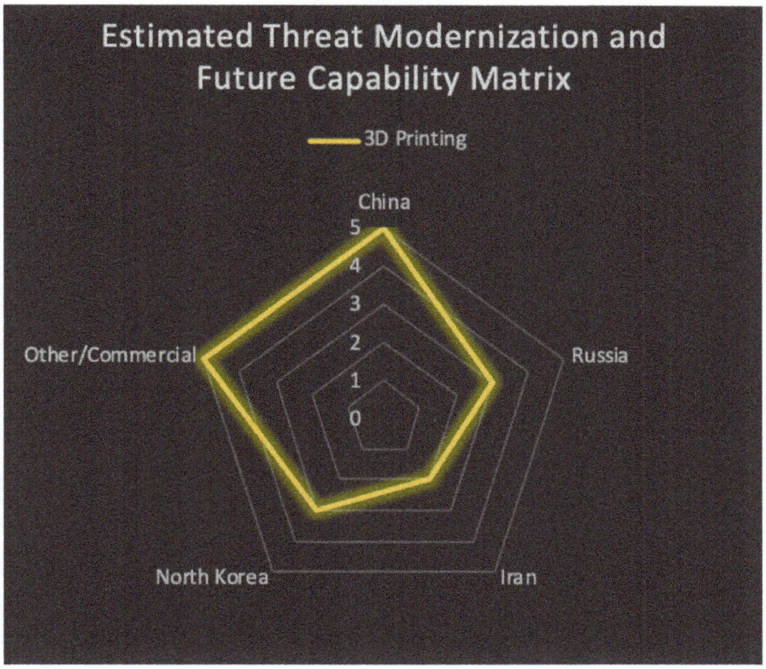

AIR DEFENSE

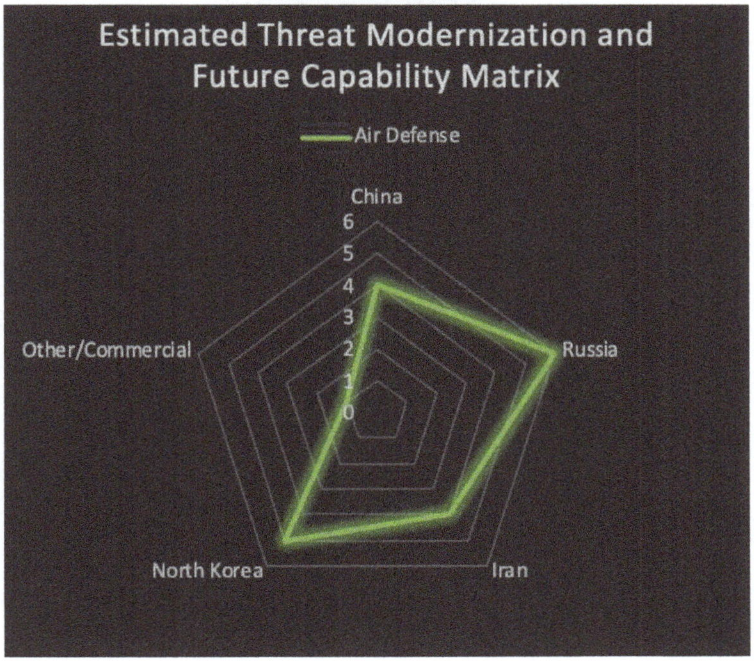

ESTIMATED TECHNOLOGY MODERNIZATION AND FUTURE CAPABILITY COMPARISON PER PLATFORM (Cont'd)

Outer rim represents higher likeliness of capability and signature being more difficult to detect by the US in 10 to 20 Years

ANTI-SATELITE

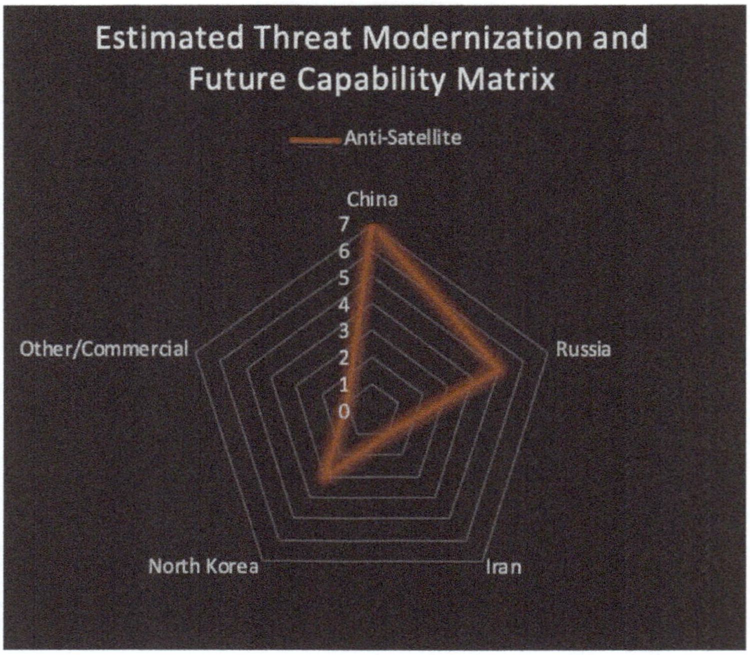

ARTIFICIAL INTELLIGENCE

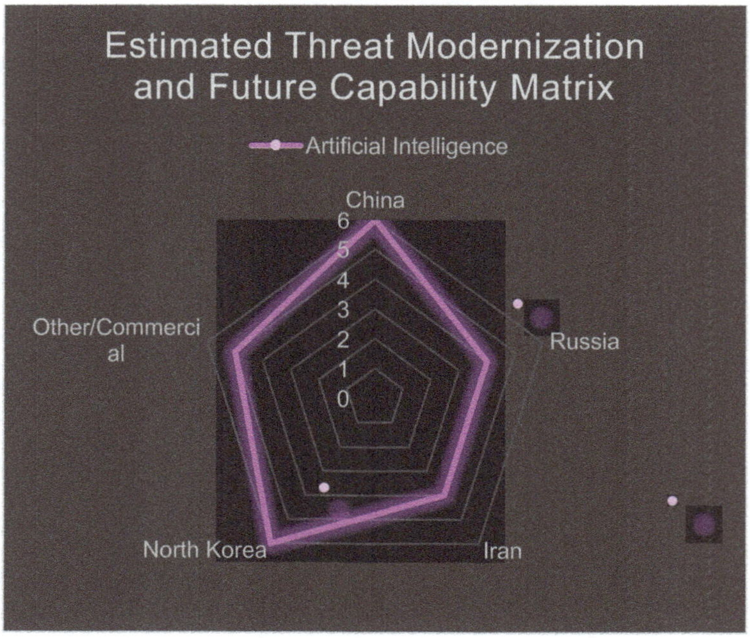

ESTIMATED TECHNOLOGY MODERNIZATION AND FUTURE CAPABILITY COMPARISON PER PLATFORM (Cont'd)

Outer rim represents higher likeliness of capability and signature being more difficult to detect by the US in 10 to 20 Years

AUTONOMOUS

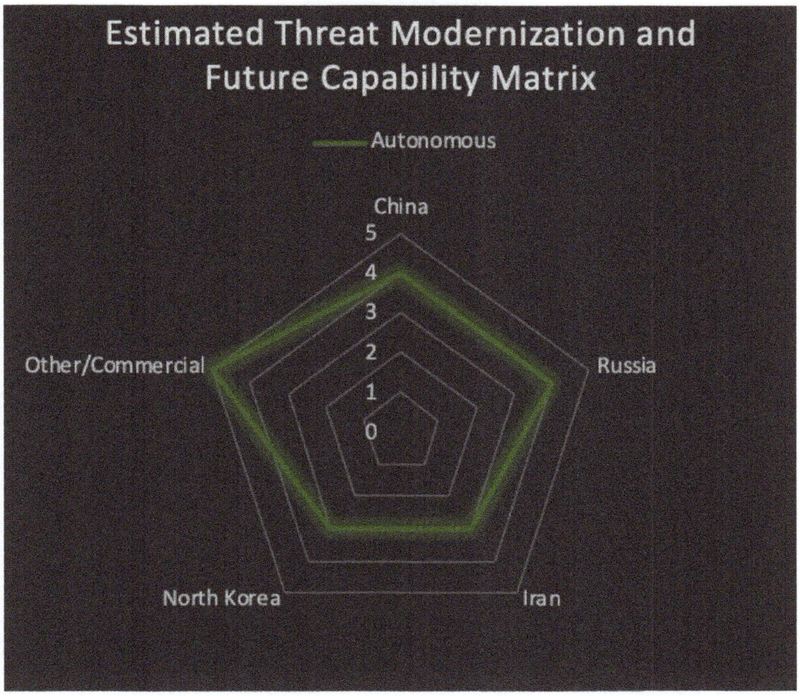

BATTERIES

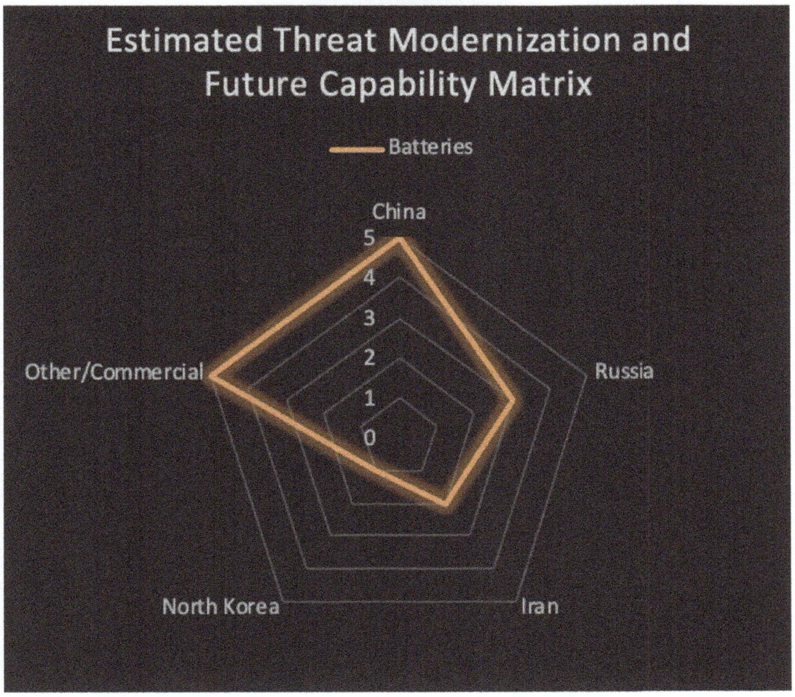

ESTIMATED TECHNOLOGY MODERNIZATION AND FUTURE CAPABILITY COMPARISON PER PLATFORM (Cont'd)

Outer rim represents higher likeliness of capability and signature being more difficult to detect by the US in 10 to 20 Years

C4ISR

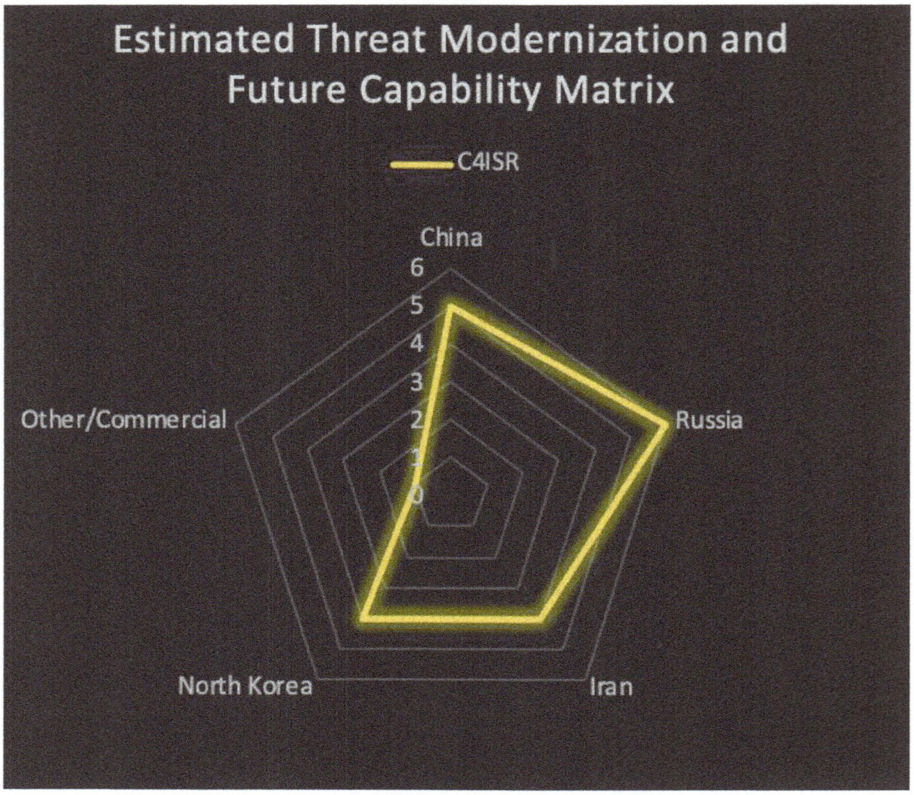

CHEMICAL

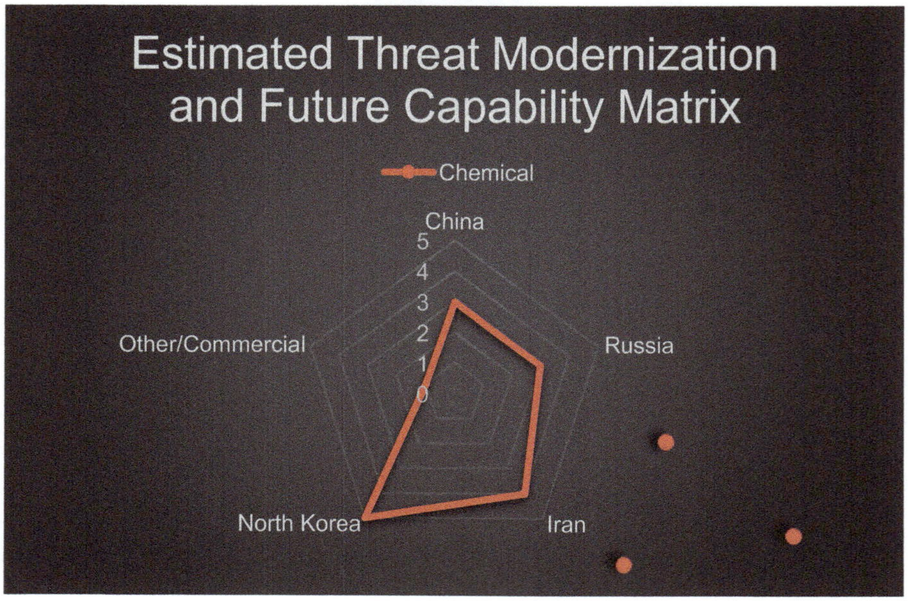

ESTIMATED TECHNOLOGY MODERNIZATION AND FUTURE CAPABILITY COMPARISON PER PLATFORM (Cont'd)

Outer rim represents higher likeliness of capability and signature being more difficult to detect by the US in 10 to 20 Years

CYBER

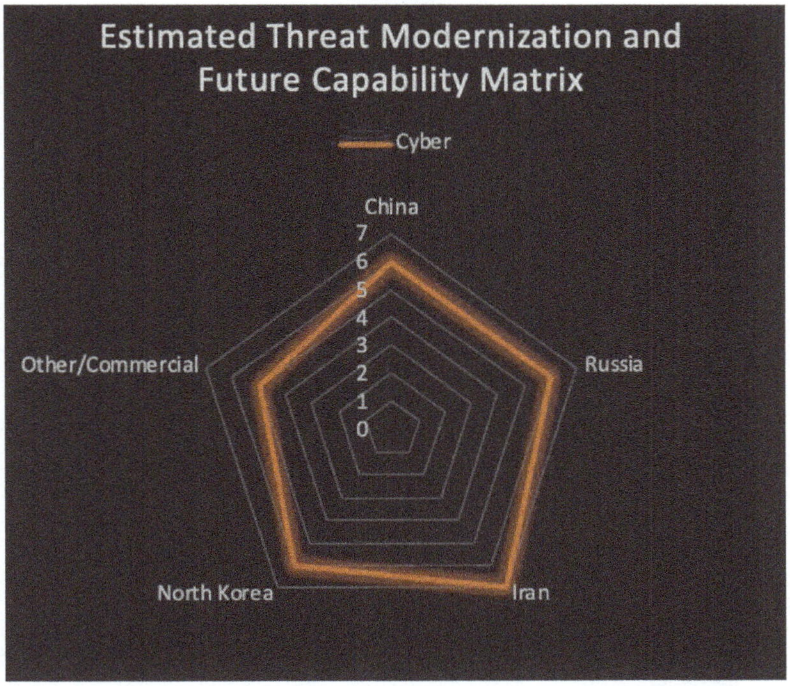

DECEPTION

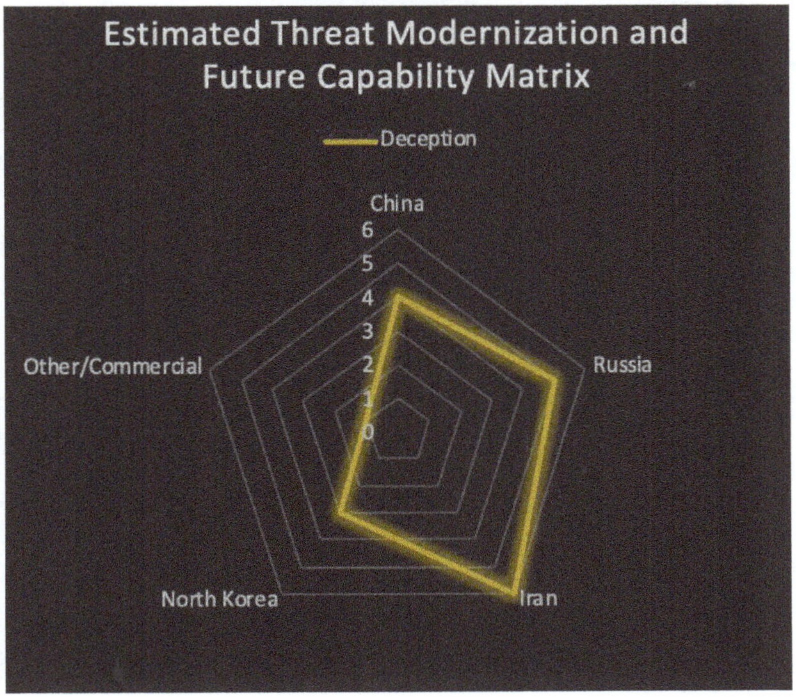

ESTIMATED TECHNOLOGY MODERNIZATION AND FUTURE CAPABILITY COMPARISON PER PLATFORM (Cont'd)

Outer rim represents higher likeliness of capability and signature being more difficult to detect by the US in 10 to 20 Years

ELECTRONIC WARFARE

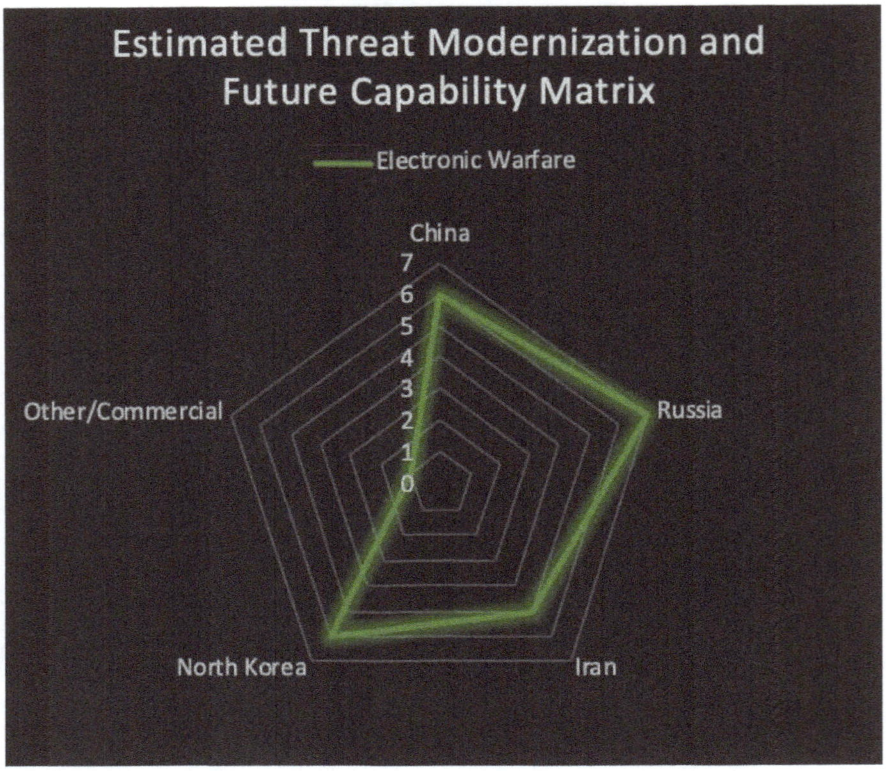

HYPERSONIC

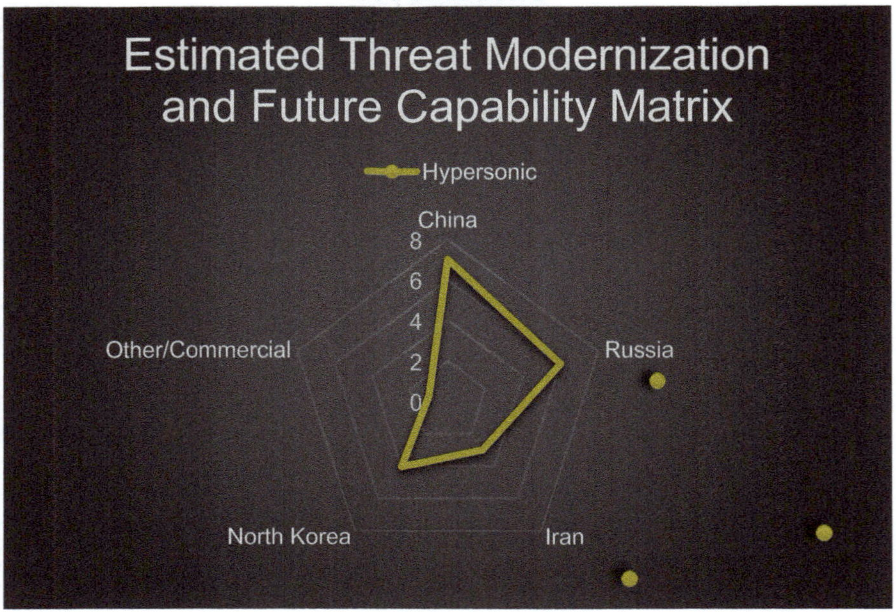

ESTIMATED TECHNOLOGY MODERNIZATION AND FUTURE CAPABILITY COMPARISON PER PLATFORM (Cont'd)

Outer rim represents higher likeliness of capability and signature being more difficult to detect by the US in 10 to 20 Years

LONG-RANGE STRIKE

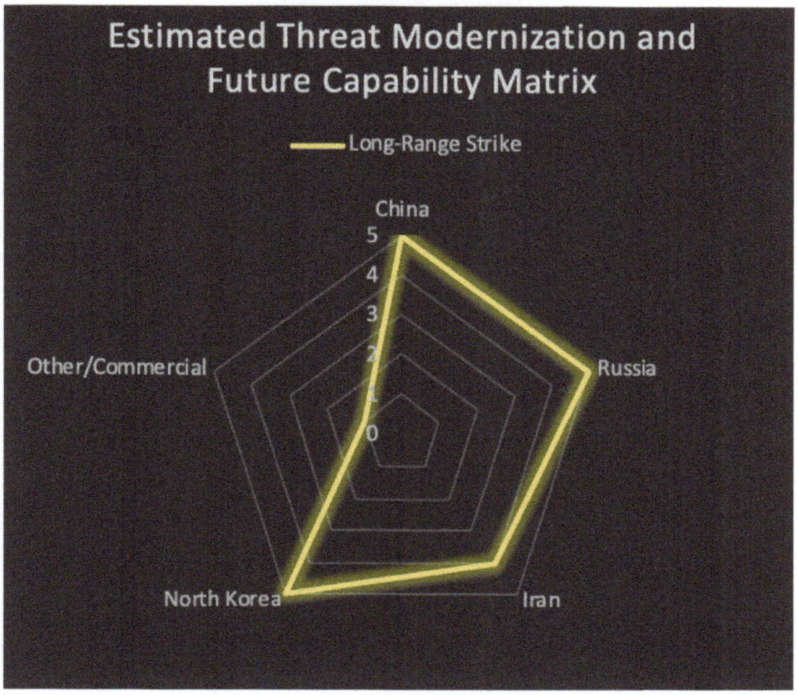

MISSILES

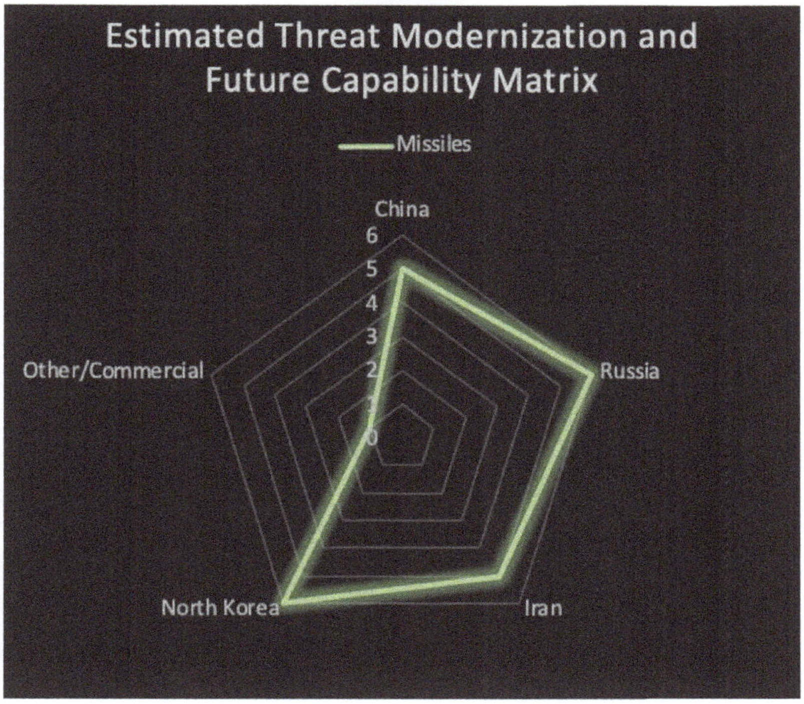

ESTIMATED TECHNOLOGY MODERNIZATION AND FUTURE CAPABILITY COMPARISON PER PLATFORM (Cont'd)

Outer rim represents higher likeliness of capability and signature being more difficult to detect by the US in 10 to 20 Years

QUANTUM

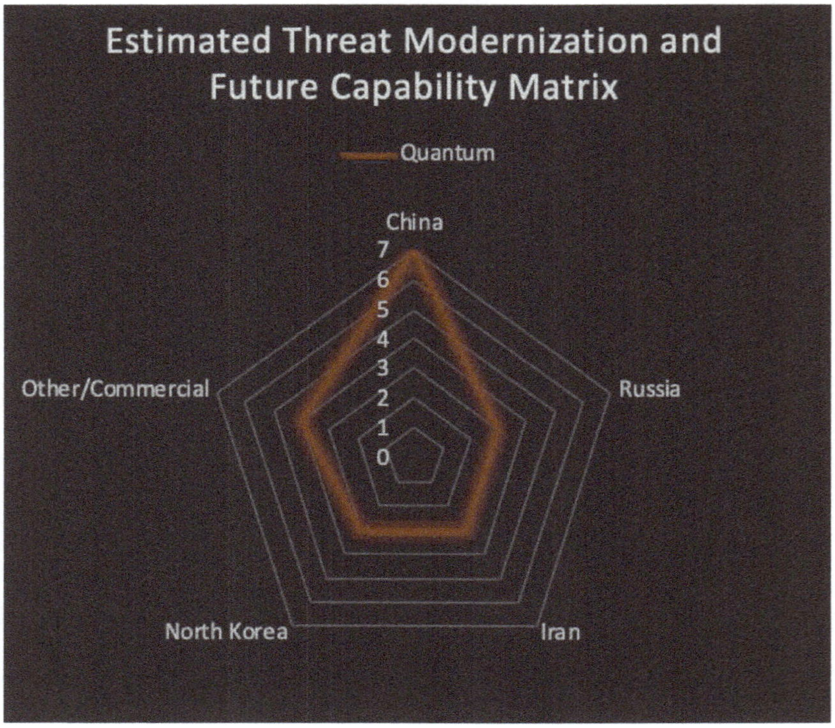

SHORT-RANGE STRIKE

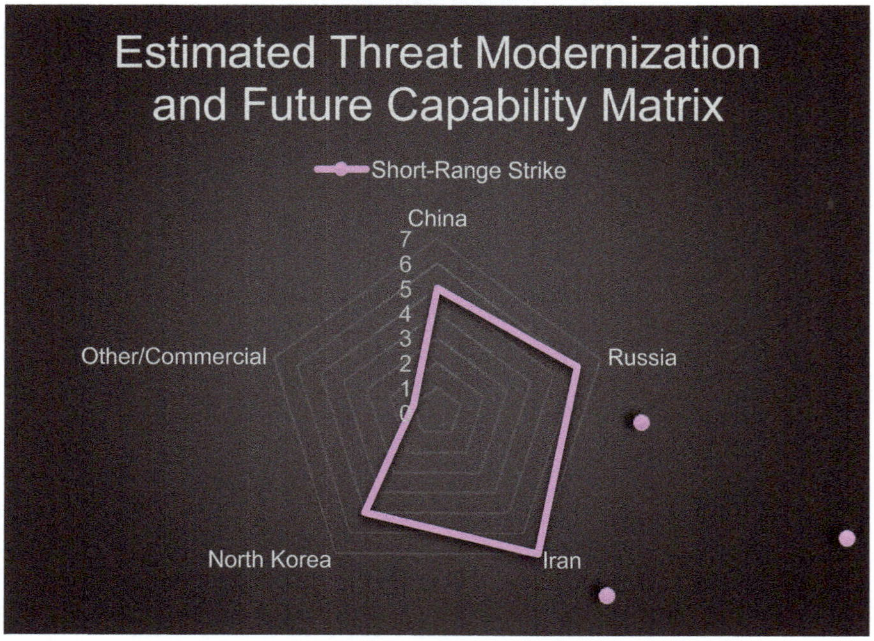

ESTIMATED TECHNOLOGY MODERNIZATION AND FUTURE CAPABILITY COMPARISON PER PLATFORM (Cont'd)

Outer rim represents higher likeliness of capability and signature being more difficult to detect by the US in 10 to 20 Years

TANKS

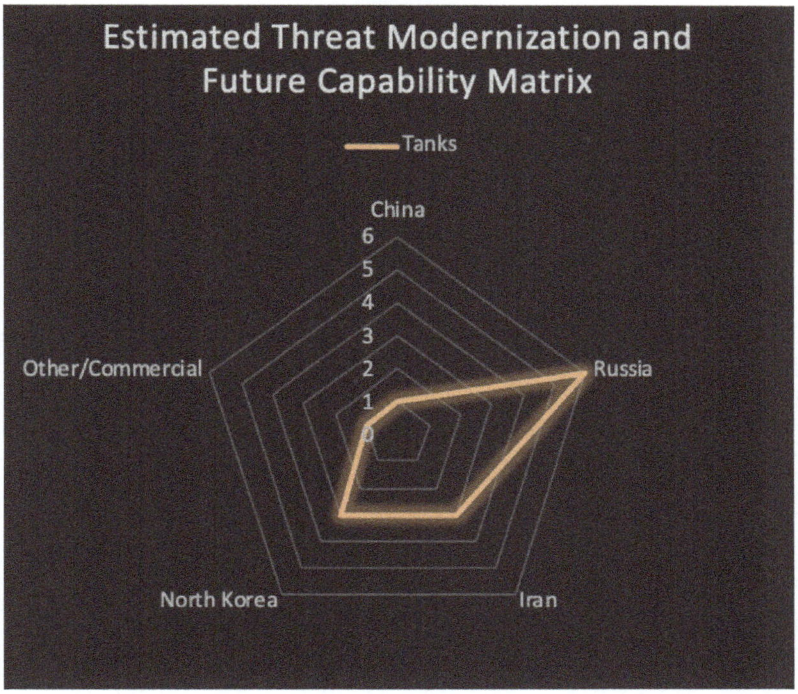

UNMANNED SYSTEMS

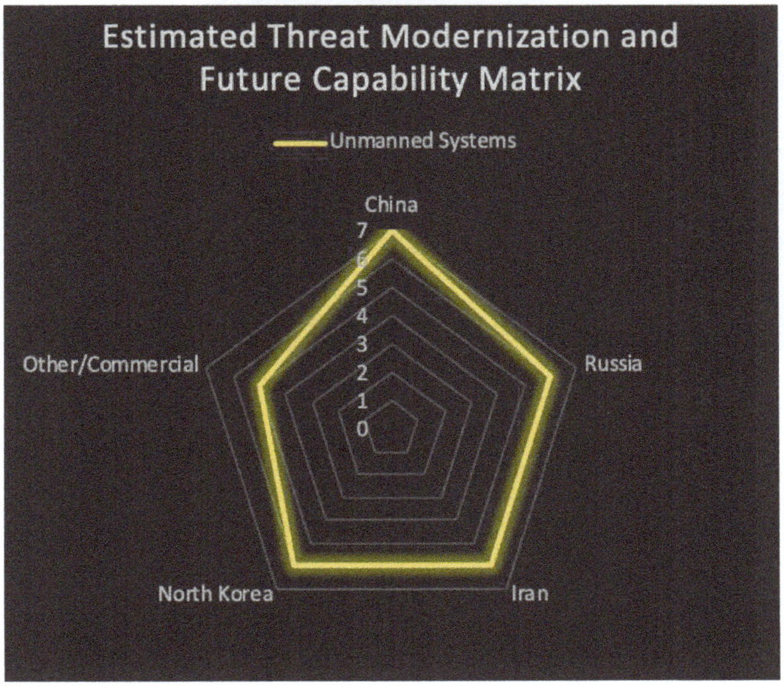

COMPARISON FOR TOP SEVEN FUTURE CAPABILITIES EFFECTED AT THE SIGNATUR/COUNTRY LEVEL

(Prior to Aggregate Cutoff)

Outer rim represents higher likeliness of capability and signature being more difficult to detect by the US in 10 to 20 Years

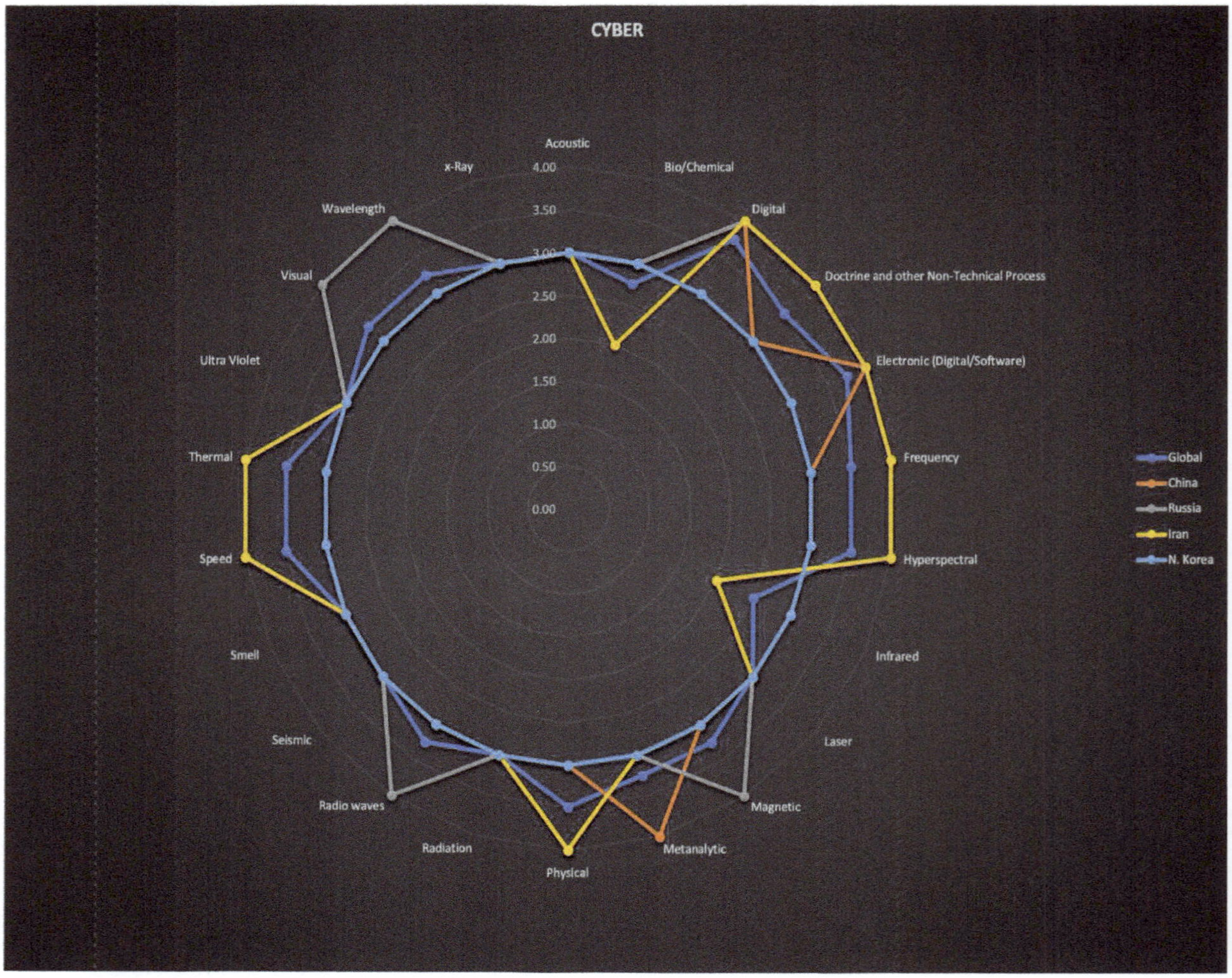

GLOBAL/COUNTRY COMPARISON FOR TOP SEVEN FUTURE CAPABILITIES EFFECTED AT THE SIGNATUR LEVEL (Cont'd)

(Prior to Aggregate Cutoff)

Outer rim represents higher likeliness of capability and signature being more difficult to detect by the US in 10 to 20 Years

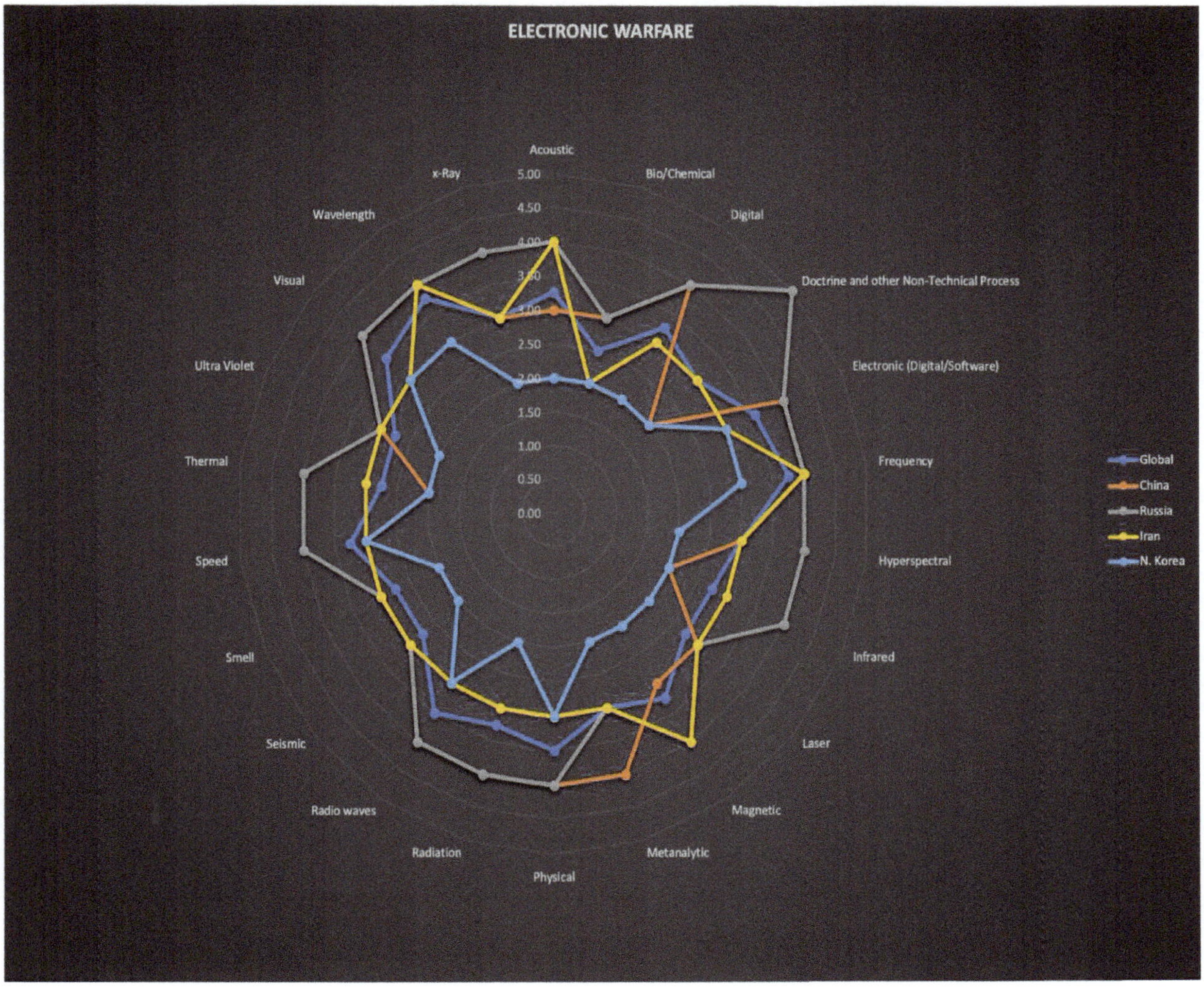

GLOBAL/COUNTRY COMPARISON FOR TOP SEVEN FUTURE CAPABILITIES EFFECTED AT THE SIGNATUR LEVEL (Cont'd)

(Prior to Aggregate Cutoff)

Outer rim represents higher likeliness of capability and signature being more difficult to detect by the US in 10 to 20 Years

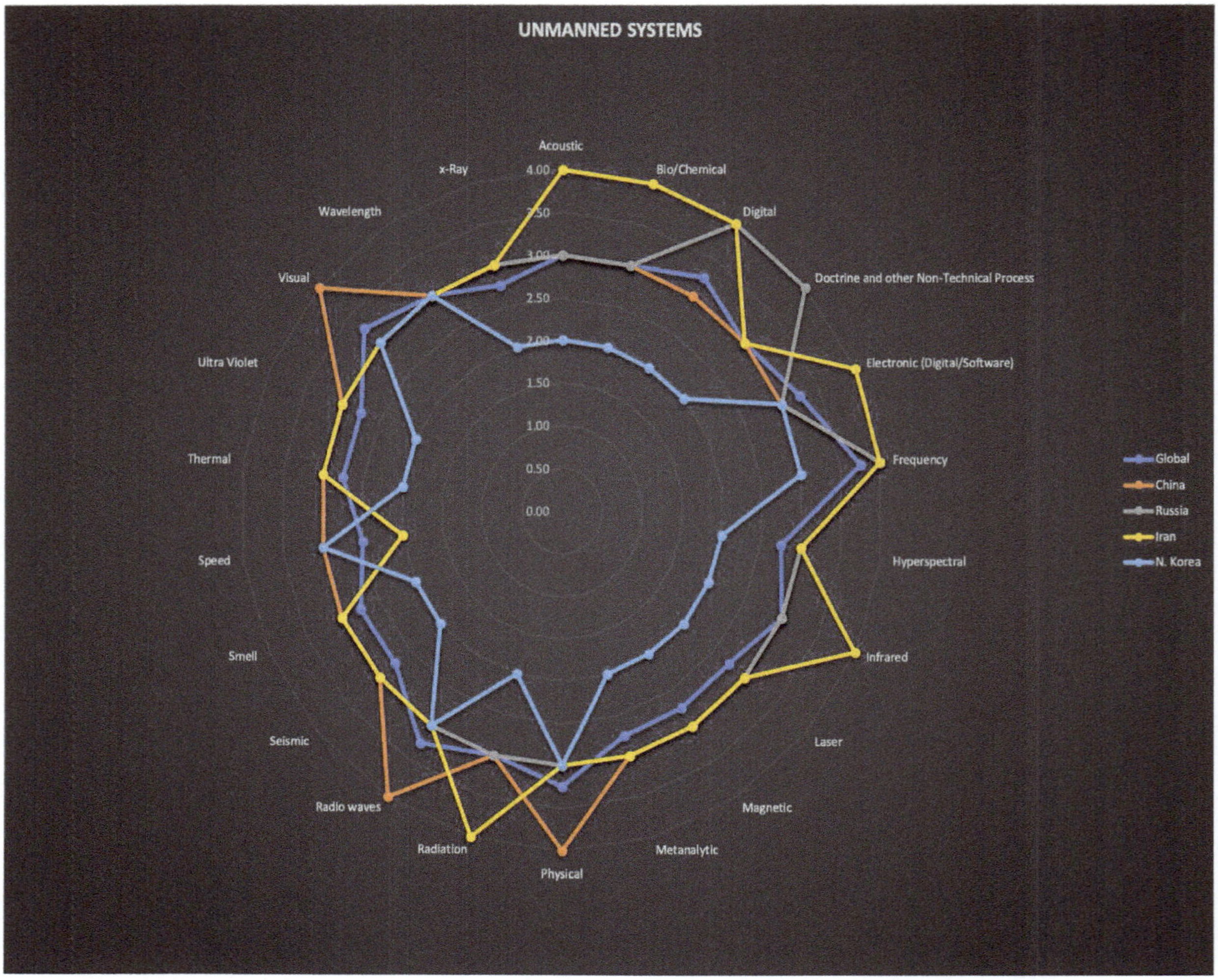

GLOBAL/COUNTRY COMPARISON FOR TOP SEVEN FUTURE CAPABILITIES EFFECTED AT THE SIGNATUR LEVEL (Cont'd)

(Prior to Aggregate Cutoff)

Outer rim represents higher likeliness of capability and signature being more difficult to detect by the US in 10 to 20 Years

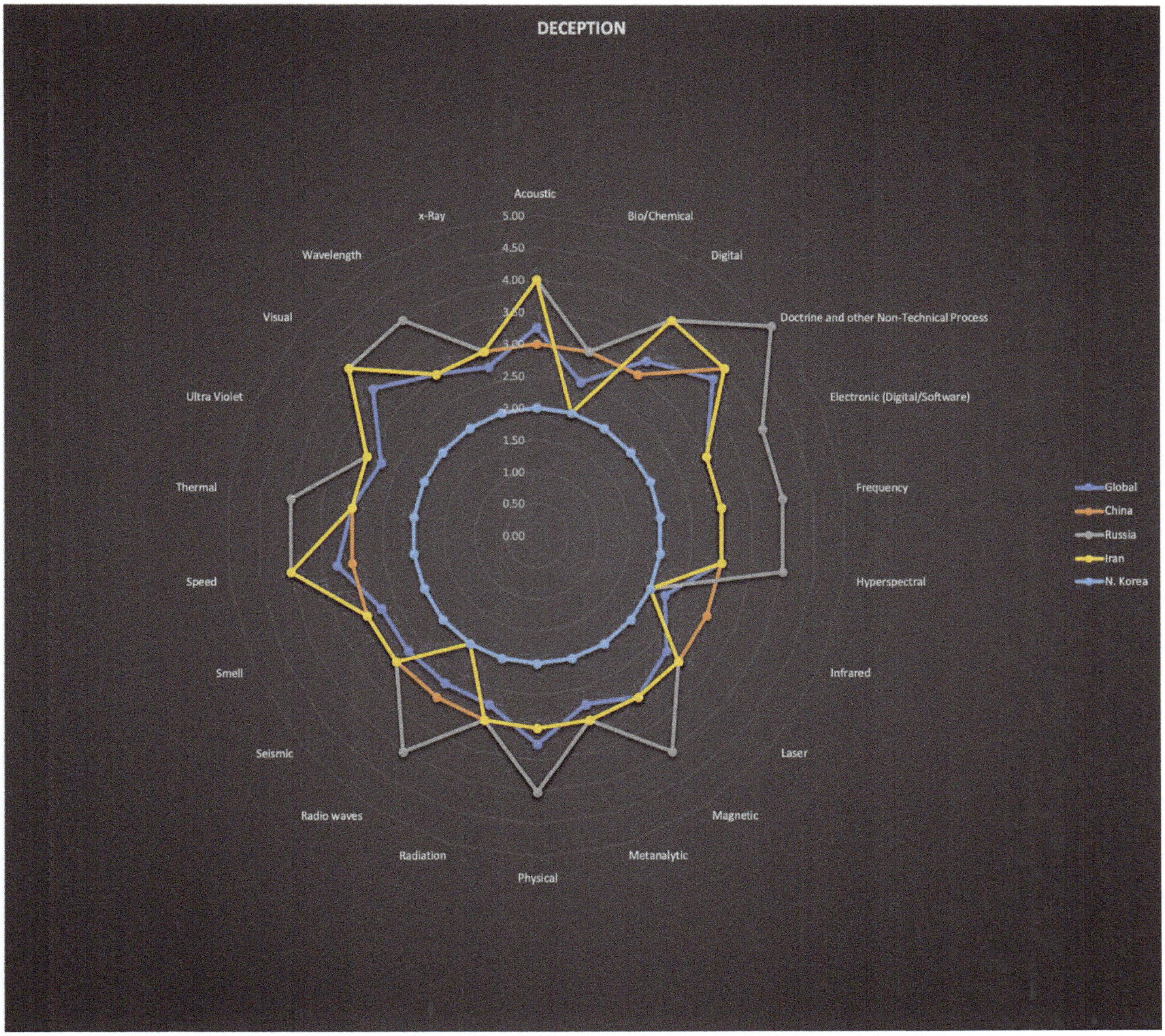

GLOBAL/COUNTRY COMPARISON FOR TOP SEVEN FUTURE CAPABILITIES EFFECTED AT THE SIGNATUR LEVEL (Cont'd)

(Prior to Aggregate Cutoff)

Outer rim represents higher likeliness of capability and signature being more difficult to detect by the US in 10 to 20 Years

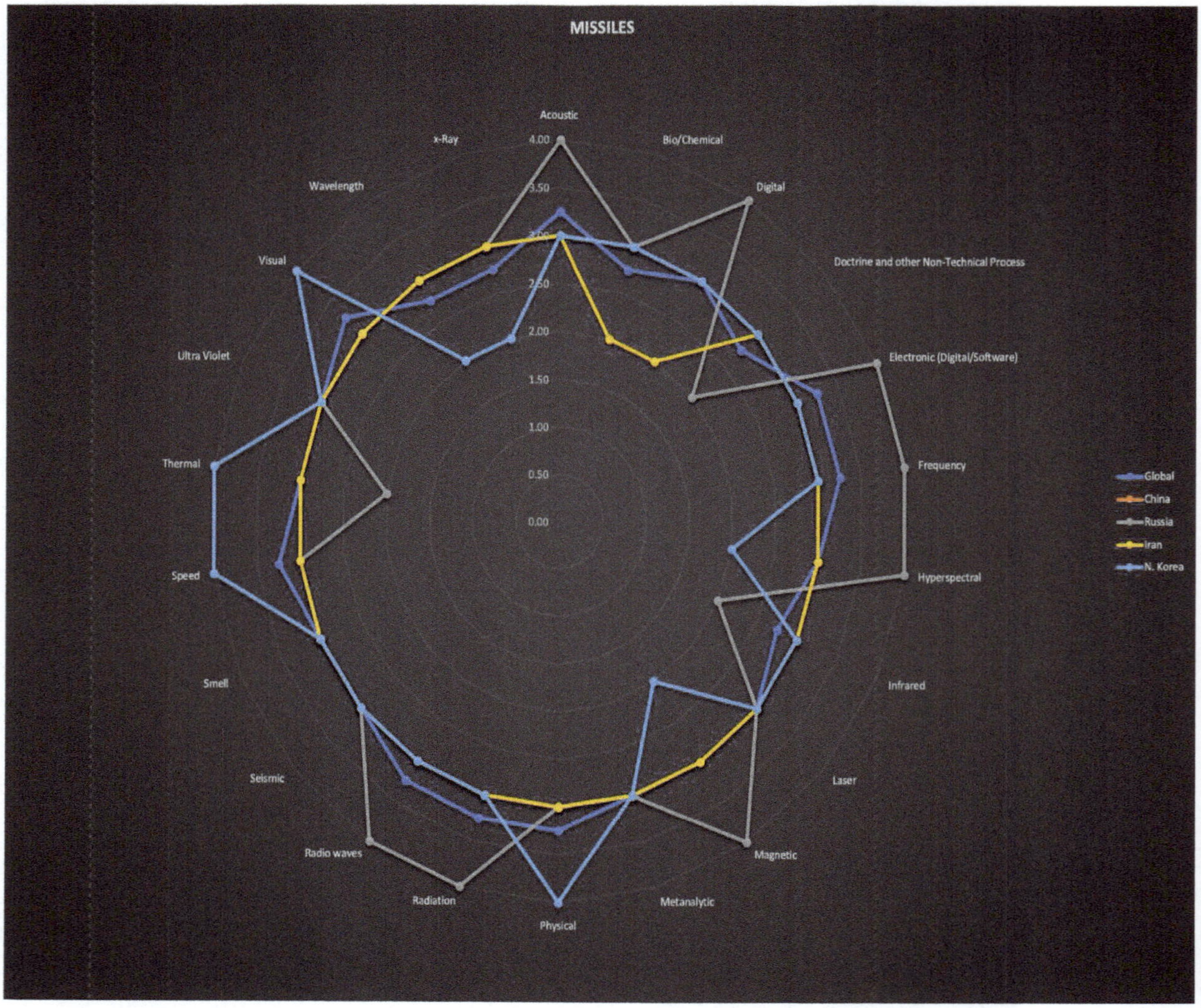

GLOBAL/COUNTRY COMPARISON FOR TOP SEVEN FUTURE CAPABILITIES EFFECTED AT THE SIGNATUR LEVEL (Cont'd)

(Prior to Aggregate Cutoff)

Outer rim represents higher likeliness of capability and signature being more difficult to detect by the US in 10 to 20 Years

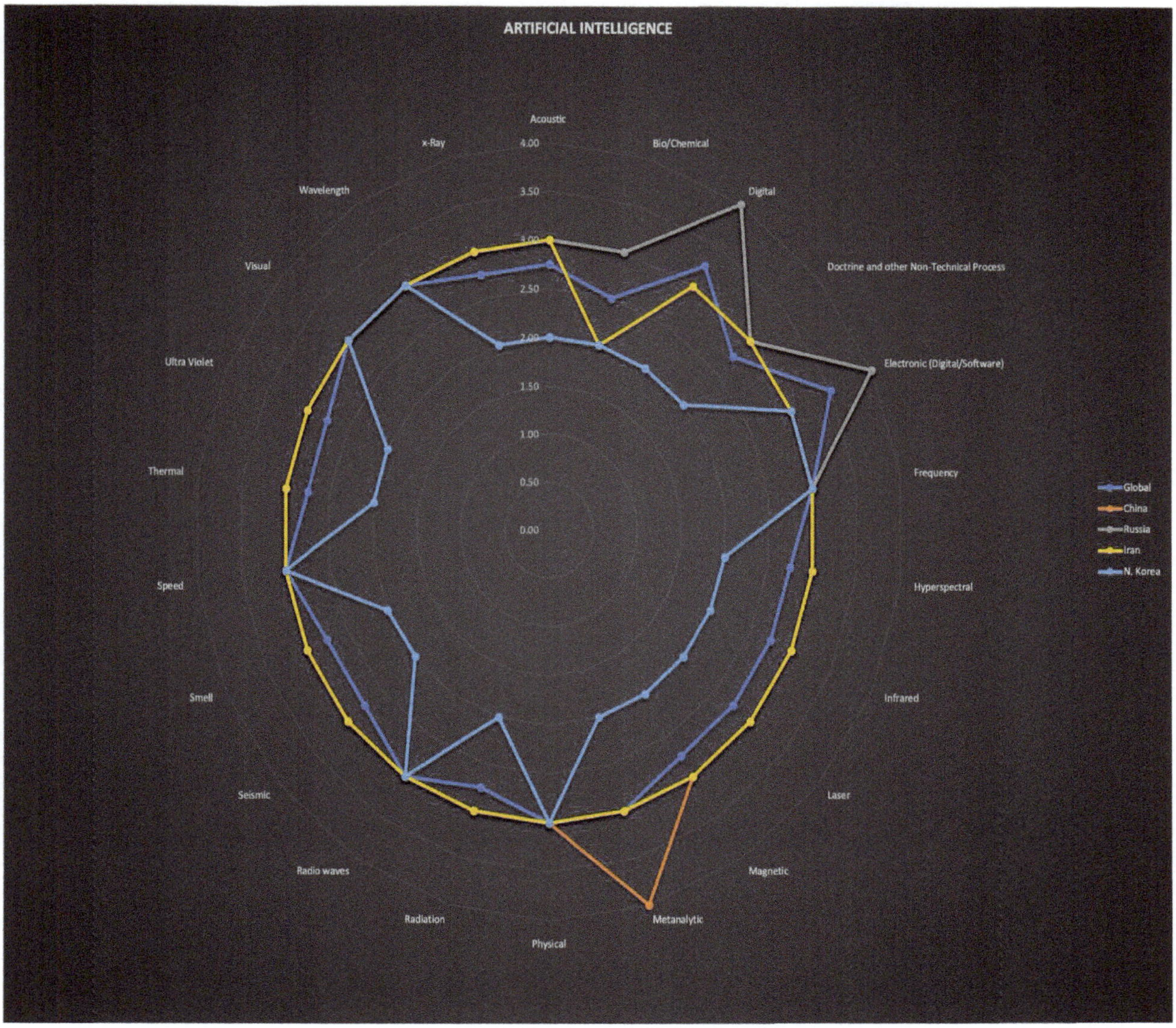

GLOBAL/COUNTRY COMPARISON FOR TOP SEVEN FUTURE CAPABILITIES EFFECTED AT THE SIGNATUR LEVEL (Cont'd)

(Prior to Aggregate Cutoff)

Outer rim represents higher likeliness of capability and signature being more difficult to detect by the US in 10 to 20 Years

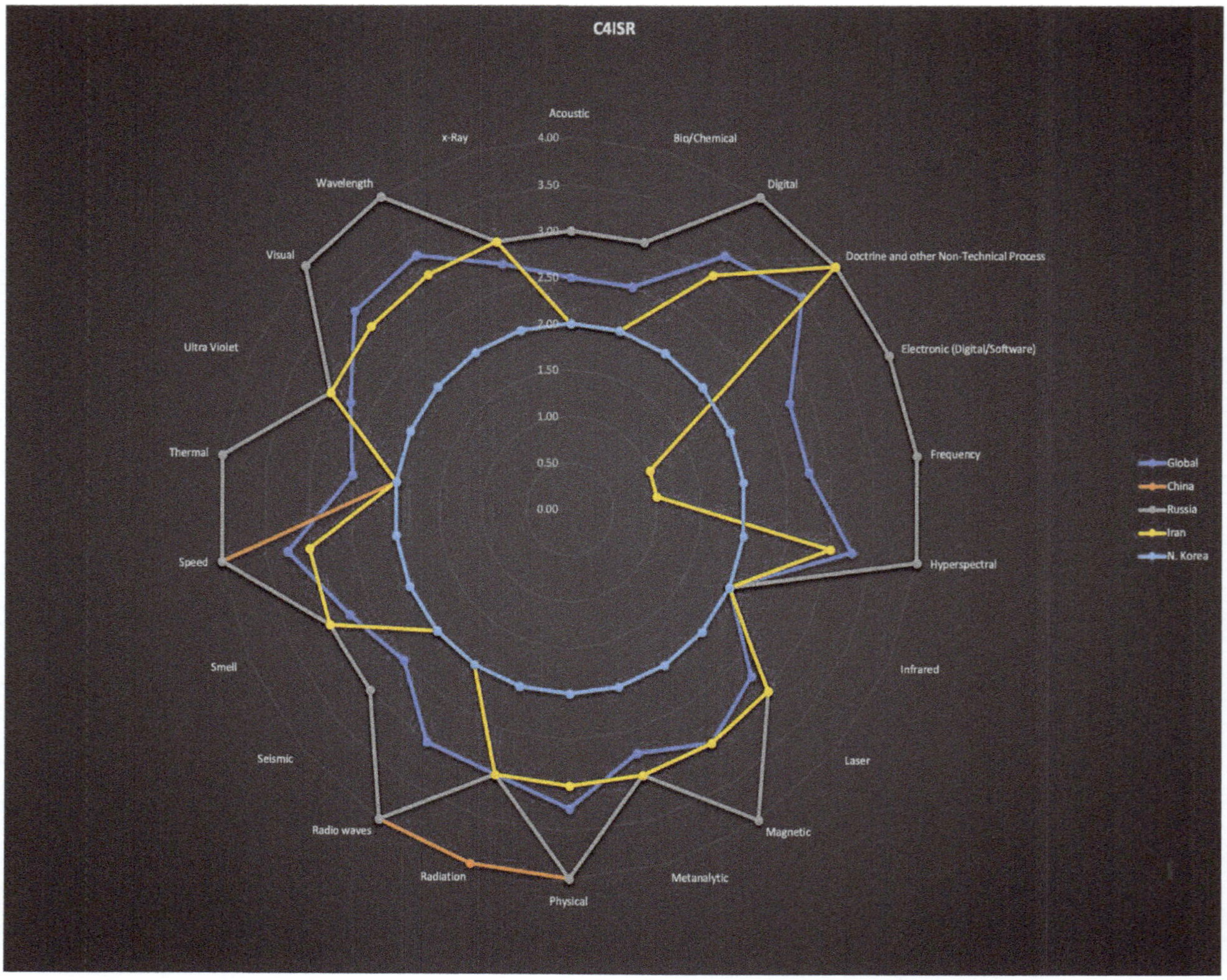

ANNEX

Annex 1
STANDARD PRIMARY SOURCE CREDIBILITY SCALE

Standard Primary Source Credibility Scale *("The Paul Scale")*			
Importance	**Factor**	**Description**	**Satisfies Criteria (Yes /No)**
HIGH	Has a good track record	Source has consistently provided true and correct information in the past	
	Information can be corroborated with other sources	Information provided by the source corroborates with information from other primary and/or secondary sources	
	Information provided is plausible	High probability of the information being true based on the analyst's experience of the topic/subject being investigated	
	Information is consistent and logically sound	Information provided is consistent when queried from different angles and is logically sound	
	Perceived expertise on the subject	Source is perceived to be an expert on the subject / topic being investigated and/or is in a role where subject knowledge is likely to be high	
	Proximity to the information	Source is close to the information – a direct participant or a witness to the event being investigated	
	Perceived trustworthiness	Source is perceived to be truthful and having integrity	
MEDIUM	No perceived bias or vested interest in the subject / topic being investigated or on the outcome of the research	Source has no perceived bias or vested interest in the subject / topic being investigated or on the outcome of the research	
	Provides complete, specific and detailed information	Information provided is specific, detailed and not generic	
LOW	Is articulate, coherent and has a positive body language	Source is articulate, coherent, has a positive body language and does not display nervousness or body language that can be construed to be evocative of deceptive behavior	
	Recommended by another trusted / credible third party	Source is recommended by others the analyst trusts but the analyst herself does not have any direct experience working with the source	
	Sociable	Source comes across as outgoing and friendly. Easy to get along with and talk to	
	Perceived goodwill to the receiver	Perceived intent or desire to help the receiver or the analyst	

Annex 2
ELECTRONIC SOURCE RELIABILITY TOOL

Trust Scale and Web Site Evaluation Worksheet (Updated OCT 2013)

Piece of Evidence #:												Score:	Trust Scale:
Criteria	**Tips**	**Value**	Y or N	Y or N	Y or N	Y or N	Y or N	Y or N	Y or N	Y or N	Y or N	0	**15-20 High**
Content can be corroborated?	Check some of the site's facts	2											**11-15 Moderate**
Recommended by subject matter expert?	Doctor, biologist, country expert	2											**6-10 Low**
Author is reputable?	Google for opinions, ask others	2											**5-0 Not Credible**
You perceive site as accurate?	Check with other sources; check affiliations	1.5											
Information was reviewed by an editor or peers?	Science journals, newspapers	1.5											
Author is associated with a reputable org?	Google for opinions, ask others.	1.5											
Publisher is reputable?	Google for opinions, ask others.	1.5											
Authors and sources identified?	Trustworthy sources want to be known	1											
You perceive site as current?	Last update?	1											
Several other Web sites link to this one?	Sites only link to other sites they trust	1											
Recommended by a generalist?	Librarian, researcher	1											
Recommended by an independent subject guide?	A travel journal may suggest sites	1											
Domain includes a trademark name?	Trademark owners protect their marks	1											
Site's bias in clear?	Bias is OK if not hidden	1											
Site has professional look?	It should look like someone cares	1											
Total		**20**											

19 Dec 2001: The criteria and weighted values are based on a survey input from 66 analysts. For details see: http://daxnorman.googlepages.com/analysis. Edited for simplicity by Kristan J. Wheaton, OCT 2013
3 Feb 2012: Excel Spreadsheet which adds auto-sum was produced by Bill Welch, Deputy Director, Center for Intelligence Research Analysis and Training, Mercyhurst College.
26 Jan 2013: Trust Scale and Web Site Evaluation Worksheet is in the PUBLIC DOMAIN.

Annex 2
ELECTRONIC SOURCE RELIABILITY TOOL (Con't)

As mentioned on page 1, Source reliability is noted at the end of each citation as **low**, **moderate**, or **high**. The citation is hyperlinked to the source and can be leveraged by clicking the superset L/M/H, you must have a valid and open internet connection for the link to work. If the source is a paid subscription, a footnote is provided at the end of each writing illustrating the source for credibility, as which time you are more than welcome to obtain an account and further read the source. Source reliability was determined using the Trust Scale and Website Evaluation Worksheet found in Annex 2 above.

Annex 3
KESSELMAN LIST OF ESTIMATIVE WORDS

The

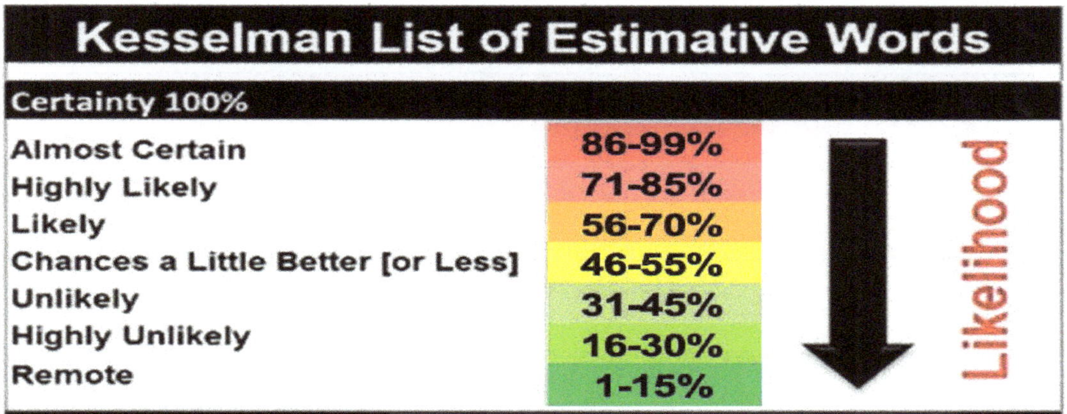

Team Sensing Version w/ Inverted Color Scale

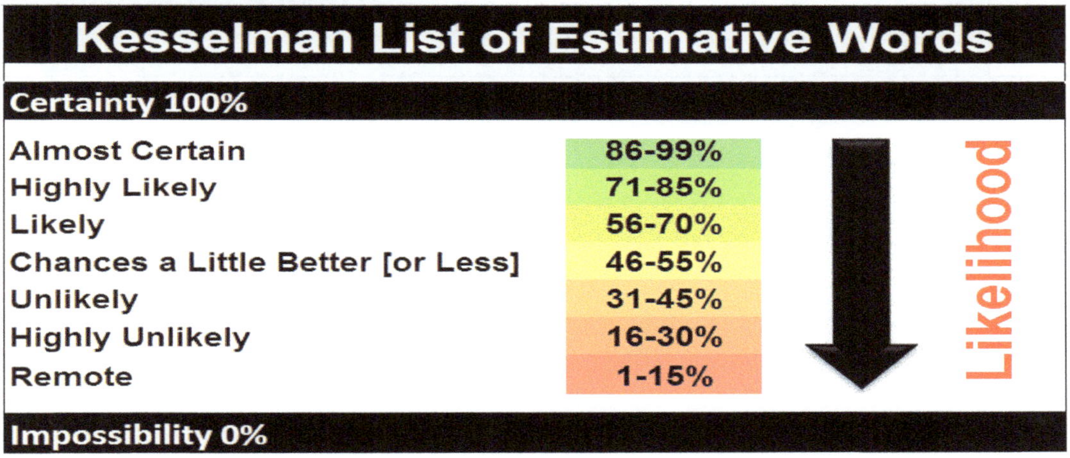

Kesselman List of Estimative Words "builds on Sherman Kent's original Words of Estimative Probability (WEP) list in the 1960s and the National Intelligence Council's current list as well as draws from Mercyhurst College's WEP list." [H] This "scale includes seven words of estimative probability which is in line with what Kent and the [National Intelligence Council] (NIC) have proposed; however, it differs in its phraseology and odds equivalents...[t]he percentile ranges are broken down into groups of 15%, except for the middle range of *chances a little better [or less]* which was assigned only 10% and the upper and lower ranges which number 14%. Absolute certainty or impossibility generally is not conveyed in intelligence assessments, but the two extremes are represented at the top and bottom of the new scale." [H] *Note: Team Sensing inverted the color scheme to illustrate uniqueness, but leveraged the Kessselman List for all other aspects.*

Annex 5
SENSING TEAM LIST OF ESTIMATIVE CHARTS PROBABILTY FOR DETECTING SIGNATURES FROM 2030 - 2040

5	Much More Difficult to Detect
4	More Difficult to Detect
3	Moderate to Routine Improvement
2	Easier to Detect
1	Extremely Easier to Detect

Annex 6
PROJECTED ADVANCED IN KEY DEPLOYABLE TECHNOLOGIES, 2020 – 2040 ACCORDING TO BROOKINGS INSTITUTE

	Moderate	High	Revolutionary
Sensors			
Chemical sensors		X	
Biological sensors		X	
Optical, Infrared, and UV sensors	X		
Radar and radio sensors	X		
Sound, sonar, and motion sensors	X		
Magnetic detection	X		
Particle beams (as sensors)	X		
Computers and communications			
Computer hardware			X
Computer software			X
Offensive cyber operations			X
System of systems/Internet of things			X
Radio communications	X		
Laser communications		X	
Artificial Intelligence/Big data			X
Quantum computer		X	
Projectiles, propulsion, and platforms			
Robotics and autonomous systems			X
Missiles	X		
Explosives		X	
Fuels	X		
Jet engines	X		
Internal-combustion engines	X		
Battery-powered engines		X	
Rockets		X	

Annex 6 (cont'd)
PROJECTED ADVANCED IN KEY DEPLOYABLE TECHNOLOGIES, 2020 – 2040 (cont'd)[31]

	Moderate	High	Revolutionary
Projectiles, propulsion, and platforms (cont'd)			
Ships	X		
Armor		X	
Stealth		X	
Satellites		X	
Other weapons and key technologies			
Radio-frequency weapons	X		
Nonlethal weapons		X	
Biological weapons		X	
Chemical weapons		X	
Other weapons of mass destruction	X		
Particle beams (as weapons)	X		
Electric guns, rail guns		X	
Lasers		X	
Nanomaterials		X	
3D printing/Additive manufacturing		X	
Human enhancement devices and substances		X	

Note: 1) The terms moderate, high and revolutionary are subjective and somewhat imprecise. In general terms, technologies showing moderate advances might improve their performance by a few percent or at most a couple of tens of percent – in terms of speed, range, lethality, or other defining characteristics – between 2020 and 2040. Those experiencing high advances will be able to accomplish tasks on the battlefield fare better than before-perhaps by 50 to 100 percent, to the extent improved performance can be so quantified. Finally, technology areas in which revolutionary advances occur will be able to accomplish important battlefield tasks that they cannot now even attempt.

---

[31] O'Hanlon, Michael. 2019. *Forecasting Change in Military Technology 2020 - 2040.* Hypothesis, Washington D.C.: Foreign Policy at Brookings.

Annex 7
ARMY G2's AGREED UPON TERMS OF REFERENCE

Terms of Reference:

Likely Threat Signatures & U.S. Army Sensor Technology in 2030-2040

For:

LTG Scott D. Berrier

By:

Team Sensing
USAWC

November 12, 2019

Terms of Reference:
Likely Threat Signatures & U.S. Army sensor technology in 2030-2040

Requirement:

What are likely future threat signatures[1] in 2030-2040? What sensors and systems[2] will the U.S. Army likely need[3] in order to detect, recognize, analyze, and target future threat signatures[4]?

Methodology:

Throughout the research project, Team Sensing will use a parallel process of modeling the research approach, collecting information/data, conducting analysis, and production of research results into oral, written & visual media.

As a precursor, the team used open source information to conduct broad, initial research on ISR and future technologies to become generally familiar with Army ISR uses and concepts. Additionally, the team finalized and implemented our information management plan and internal communication plan but may alter to best suit research needs.

The team will maintain, as appropriate, a Definitions of Terms (See ANNEX) for definition consistency.

The team will continue to use open source information to further conduct research and collect information applicable to address the primary question.

The team will consider breaking up the question into relevant sections for focused information and data collection. Possible sections could include, but not limited to:

- Emerging communication theories
- Academic communication studies
- Emerging laboratory experiment plans
- Ongoing laboratory experiments process & results
- Emerging technologies
- Applied technologies
- Case studies
- Ongoing meta-analysis
- Existing meta-analysis
- Future targeting and processing

Where appropriate, the team may use "push" (e.g., RSS feeds) and "pull" sources (e.g., internet searches and proprietary databases) to rapidly develop a comprehensive understanding of the information space surrounding sensor technology, emerging complex communication, and ISR domains.

The team will apply appropriate structured analytic techniques and methodologies that provide insight into the collected data and enhance the rigor of analysis. Techniques and methodologies will be selected based on their relative best fit with the requirement and the types of data available, and may include, but not limited to:

- Mind mapping
- Nominal method technique
- Red team analysis
- Alternative futures analysis
- Multi-criteria analysis
- Science of estimation
- Forecasting
- Contrarian techniques
- Diagnostic techniques
- Imaginative thinking techniques

The team will consider a system of internal quality controls that reviews each final piece of analysis. Standards of analysis may include, but are not limited to, logic, sourcing, correct application of analytic methods, rechecking assumptions, and quality of information checks to ensure value added.

Team Sensing will deliver a complete written report in PDF format, containing written findings, along with any visual or tactile supporting media. Delivery will include an oral briefing on the team's key findings.

Challenges:

The team identified the following as some potential challenges during the research project:

- Time. The research project must be completed by April 2020. Full academic demands at Army War College constrains project scope on research and analysis in order to produce meaningful findings.
- Funding. There are no funds currently allocated to fund this project; however, it is unlikely the research will require funds.
- TDY. With both time and funding constraints, TDY to collect information is improbable; however, the team believes TDY is not a critical aspect to conduct adequate research.
- Limited information sources. Due to time and equipment constraints, the team has access to mostly open source information and the final product will be UNCLASSIFIED.
- Futures analysis/forecasting experience. This research project is the team's first futures forecasting; however, team dedication & expert classroom instruction help mitigate lack of experience.

Resources:

The team identified the following resources available during the research project:

- Intelligence Experience. Only two of five team members have experience in the intelligence community -- one Army MI officer and one Air Force Special Agent. However, the team's diverse backgrounds, skillsets, & experiences helps mitigate perceptual and cognitive biases.
- Institutional. The USAWC Library has a wealth of proprietary databases.
- Library Researcher. The team is pleased to have Mr. Thomas Moss as the team's dedicated researcher.

Administration:

The project's end date is estimated for NLT April 2020 with delivery of findings preferably within the window of 30 March – 3 April 2020; however, this will be coordinated based on scheduled availability.

All correspondence between the research team and the decision maker will occur through the primary point of contact, LTC Sam Smith. If the primary point of contact is unavailable, all correspondence will be handled by the secondary point of contact, Mr. Ashraf Abdelhak.

Team Sensing Members:

- Mr. Ashraf Abdelhak, GS-15, AFOSI. ashraf.abdelhak@armywarcollege.edu
- COL Jerry Brown, US Army Reserves, Logistics. jerry.brown@armywarcollege.edu
- LTC Alex Duran, US Army, Infantry. rafael.duranmariot@armywarcollege.edu
- COL Russ Hoff, US Army, Acquisition Corps. russell.hoff@armywarcollege.edu
- LTC Sam Smith, US Army, Military Intelligence. samuel.smith@armywarcollege.edu

Definitions of Terms

The following definitions consist of either doctrinal definitions or the team's definitions:

- Air domain — The atmosphere, beginning at the Earth's surface, extending to the altitude where its effects upon operations become negligible. (JP 3-30)
- Analyze – Extraction of meaningful information and the conversion of processed information into intelligence through the integration, evaluation, and interpretation of all source data and the preparation of intelligence products in support of known or anticipated user requirements.

- Cyberspace — A global domain within the information environment consisting of the interdependent networks of information technology infrastructures and resident data, including the Internet, telecommunications networks, computer systems, and embedded processors and controllers. (JP 3-12)
- Detect – In tactical operations, the perception of an object of possible military interest but unconfirmed by recognition (JP 1-02)
- Full-spectrum superiority — The cumulative effect of dominance in the air, land, maritime, and space domains; electromagnetic spectrum; and information environment (which includes cyberspace) that permits the conduct of operations without effective opposition or prohibitive interference. (JP 3-0)
- ISR – intelligence, surveillance, and reconnaissance. An integrated operations and intelligence activity that synchronizes and integrates the planning and operation of sensors, assets, and processing, exploitation, and dissemination systems in direct support of current and future operations. (JP 2-01)
- Land domain – The area of the Earth's surface ending at the highwater mark and overlapping with the maritime domain in the landward segment of the littorals. (JP 3-31)
- Recognize – Confirmation of military interest in which the signal can be distinguished by a category.
- Sensor – A device, module, machine, or subsystem whose purpose is to detect events or changes in its environment and send the information to other electronics, devices, modules, or subsystems.
- Signature – distinctive features or characteristics of fixed or dynamic target sources
- Space domain — The area above the altitude where atmospheric effects on airborne objects become negligible. (JP 3-14)
- Target – The detection, identification, and location of a target in sufficient detail to permit the effective employment of capabilities that create the required effects. Also, Target acquisition. (JP 3-60)

Annex 8
DEFENITIONS AND TERMS USED

Acoustic - Acoustic Intelligence — Intelligence derived from the collection and processing of acoustic phenomena. (JP 2-0) Of or relating to the sense or organs of hearing, to sound, or to the science of sounds (Source: Merriam Dictionary)
Anti-Satellite - System designed for the destruction or incapacitation of satellites
Artificial Intelligence - A branch of computer science dealing with the simulation of intelligent behavior in computers; the capability of a machine to imitate intelligent human behavior
Batteries - A container consisting of one or more cells, in which chemical energy is converted into electricity and used as a source of power.
Ballistic Missile (BM) - is any missile that does not rely upon aerodynamic surfaces to produce lift and follows a ballistic trajectory when thrust is terminated.
Bio/Chemical - chemical, biological, radiological, and nuclear defense – (DOD) Measures taken to minimize or negate the vulnerabilities and/or effects of a chemical, biological, radiological, or nuclear incident. Also called CBRN defense. (JP 3-11) See ATP 4-02.84. chemical, biological, radiological, or nuclear incident – (DOD) Any occurrence, resulting from the use of chemical, biological radiological and nuclear weapons and devices; the emergence of secondary hazards arising from counterforce targeting; or the release of toxic industrial materials into the environment, involving the emergence of chemical biological, radiological and nuclear hazards.
C4ISR - Concept of Command, Control, Communications, Computers, Intelligence, Surveillance and Reconnaissance, the U.S. term for C4STAR
Chemical - Relating to, used in, or produced by chemistry or the phenomena of chemistry
Cyber - Relating to, or involving computers or computer networks (such as the Internet)
Deception - The act of causing someone to accept as true or valid what is false or invalid Source: DOD Dictionary of Military and Associated Terms (As of January 2020)
Digital - 1) (Computers) performing internal logical and arithmetic operations by means of digits, usually represented as binary numbers. Contrasted to analog, wherein variables are represented as continuous physical quantities such as voltages or the position of a pointer on a continuous scale; as, a digital computer. (Source: Definition.net) H 2) composed of data in the form of especially binary digits, Digital Images/Photos, A digital readout
Digital Broadcast - a broadcast employing digital communications signals
providing a readout in numerical digits: a *digital voltmeter* a *digital* watch/clock
Relating to an audio recording method in which sound waves are represented digitally (as on magnetic tape) so that in the recording wow and flutter are eliminated and background noise is reduced. (Source: Merriam Webster)
Electronic Warfare Support — Division of electronic warfare involving actions tasked by, or under direct control of, an operational commander to search for, intercept, identify, and locate or localize sources of intentional and unintentional radiated electromagnetic energy for the purpose of immediate threat recognition, targeting, planning and conduct of future operations. Also called ES. See also electronic attack; electronic protection; electronic warfare. (JP 3-13.1)

Electronic Warfare. (JP 3-13.1 Joint Doctrine for Command and Control Warfare (C2W) 7 Feb 97) Any military action involving the use of electromagnetic and directed energy to control the electromagnetic spectrum or to attack the enemy. Also called EW. The three major subdivisions within electronic warfare are: electronic attack, electronic protection, and electronic warfare support.

Electronic Digital and Software - The use of electron non-analog encoded computer instructions (deduced definition from the information below).

Electronics - a branch of physics that deals with the emission, behavior, and effects of electrons (as in electron tubes and transistors) and with electronic devices) Source: Merriam Dictionary

Electronic warfare (joint) - Military action involving the use of electromagnetic and directed energy to control the electromagnetic spectrum or to attack the enemy.

Electronic warfare consists of three divisions: electronic attack, electronic protection, and electronic warfare support. (JP 3-13.1)

Electronic Intelligence — Technical and geolocation intelligence derived from foreign noncommunications electromagnetic radiations emanating from other than nuclear detonations or radioactive sources. Also called ELINT. See also electronic warfare; foreign instrumentation signals intelligence; intelligence; signals intelligence. (JP 3-13.1)

Frequency - the number of complete oscillations per second of energy (such as sound or electromagnetic radiation) in the form of waves (Source: Merriam Dictionary). In physics, the number of waves that pass a fixed point in unit time; also, the number of cycles or vibrations undergone during one unit of time by a body in periodic motion. A body in periodic motion is said to have undergone one cycle or one vibration after passing through a series of events or positions and returning to its original state. (Source: Britannica Encyclopedia).

Hyperspectral - hyperspectral imagery — Term used to describe the imagery derived from subdividing the electromagnetic spectrum into very narrow bandwidths allowing images useful in precise terrain or target analysis to be formed. Also called HSI. (JP 2-03)

Hypersonic - noting or pertaining to speed that is at least five times that of sound in the same medium. Source dictionary.com

Infrared - situated outside the visible spectrum at its red end —used of radiation having a wavelength between about 700 nanometers and 1 millimeter (Source: Merriam Webster)

Missile - an object (such as a weapon) thrown or projected usually so as to strike something at a distance (Source: Merriam Webster)

Low-altitude missile engagement zone — In air and missile defense, that airspace of defined dimensions within which the responsibility for engagement of air and missile threats normally rests with low- to medium-altitude surface-to-air missiles.

Guided missile — An unmanned vehicle moving above the surface of the Earth whose trajectory or flight path is capable of being altered by an external or internal mechanism. See also ballistic missile. (JP 3-01)

Short-Range Ballistic missile (Short-range strike). A ballistic missile with a range capability between 300-600 nautical miles. Also called SRBM. (Approved for incorporation into the DOD Dictionary.) JP 3-01

Medium-range ballistic missile. A ballistic missile with a range capability from about 600 to 1,500 nautical miles. Also called MRBM. (DOD Dictionary. SOURCE: JP 3-01)

Intermediate-range Ballistic Missile. A ballistic missile with a range capability from 1,500 to 3,000 nautical miles. Also called IRBM. (Approved for incorporation into the DOD Dictionary.)
Long-range strike missiles: A ballistic missile with a range capability beyond 3,000 nautical miles. intercontinental ballistic missile — A long-range ballistic missile with a range capability greater than 3,000 nautical miles. Also called ICBM. (JP 3-01)
Missile Defense — Defensive measures designed to destroy attacking enemy missiles, or to nullify or reduce the effectiveness of such attack. Also called MD. (JP 3-01)
Quantum: Quantity or Amount (Source: Collins Dictionary)
(Google announced a 72-qubit processor in 2018 – surpassing IBM's record the previous year of 50 qubits – and said that its new chip might achieve quantum supremacy within a year.) H
The term 'quantum supremacy' refers to the ability of a quantum computer to perform tasks beyond the capability of today's most powerful conventional supercomputers. (Source: The Military Balance 2019: Quantum Computing and Defence, Feb 19). H
Quantum Computing should best be conceived of as an alternative, complementary and even synergistic technology that will be able to solve some problems that current computers cannot, but which will most likely also be comparatively ineffective, or only marginally better, for solving other problems at which current computers excel. (Source: The Military Balance 2019: Quantum Computing and Defence, Feb 19). H
Laser - a device that utilizes the natural oscillations of atoms or molecules between energy levels for generating a beam of coherent electromagnetic radiation usually in the ultraviolet, visible, or infrared regions of the spectrum (Source: Merriam Webster)
Magnetic - any piece of certain kinds of materials, as iron, that has the property of attracting like material; this property may be permanent or temporarily induced such as electromagnet. (Source Collins Dictionary)
Metanalytic - Meta-analysis is the statistical procedure for combining data from multiple studies. When the treatment effect (or effect size) is consistent from one study to the next, meta-analysis can be used to identify this common effect. When the effect varies from one study to the next, meta-analysis may be used to identify the reason for the variation. (Source: Comprehensive metal-analysis). H
Radiation - The process of emitting radiant energy in the form of waves or particles. (Source: Merriam Webster)
Radio Waves - A radio wave is a moving electromagnetic field that has velocity in the direction of travel. Its components are of electric and magnetic intensity arranged at right angles to each other. Once a wire is connected to a transmitter and properly grounded, it begins to oscillate electrically, causing the wave to convert nearly all of the transmitter power into an electromagnetic radio wave. (Source: FM 6-02.53. Tactical Radio Operations. 5 August 2009.)
Smell - to perceive the odor or scent of through stimuli affecting the olfactory nerves**:** get the odor or scent of with the nose
Software - Encoded computer instructions, usually modifiable (unless stored in some form of unaltered memory such as ROM)
Physical characteristics — Those military characteristics of equipment that are primarily physical in nature. (JP 3-60)

Seismic - of or relating to an earth vibration caused by something else (such as an explosion or the impact of a meteorite) (Source: Merriam Webster
Speed the magnitude of movement irrespective of direction (Source: Merriam Webster).
Stealth - The stealth operations are supported, for example, in the U.S. Army's field manuals FM 90-2 (1988) and FM 20-3 (1999) as camouflage, concealment, and decoys (CCD) tactics to hide, disguise, decoy, or disrupt the appearance of military targets.
An aircraft-design characteristic consisting of oblique angular construction and avoidance of vertical surfaces that is intended to produce a very weak radar return —usually used before another noun (source: Merriam Webster)
Tank - an enclosed armored military vehicle; has a cannon and moves on caterpillar treads (source: vocabulary.com)
Thermal - Means relating to or caused by heat or by changes in temperature. (Source: Collins Dictionary).
Thermal Radiation - process by which energy, in the form of electromagnetic radiation, is emitted by a heated surface in all directions and travels to its point of absorption at the speed of light; thermal radiation does not require an intervening medium to carry it. (Source: Encyclopedia Britannica).
Thermal Imaging - photography that captures 'heat pictures' rather than ordinary light pictures of objects. (Source: Merriam Webster)
Ultraviolet - situated beyond the visible spectrum at its violet end —used of radiation having a wavelength shorter than wavelengths of visible light and longer than those of X-rays (Source: Merriam Webster)
Unmanned systems: unmanned aircraft — An aircraft that does not carry a human operator and is capable of flight with or without human remote control. Also called UA. (JP 3-30; DOD Dictionary of Military and Associated Terms)
Unmanned Aircraft System — That system whose components include the necessary equipment, network, and personnel to control an unmanned aircraft. Also called UAS. (JP 3-30; DOD Dictionary of Military and Associated Terms) ADP 2.0
Visual: — Various visual media with or without sound that generally includes still and motion photography, audio video recording, graphic arts, and visual presentations.
Wavelength: the distance in the line of advance of a wave from any one point to the next point of corresponding phase (Source: Merriam Webster)
X-Ray: any of the electromagnetic radiations that have an extremely short wavelength of less than 100 angstroms and have the properties of penetrating various thicknesses of all solids, of producing secondary radiations by impinging on material bodies, and of acting on photographic films and plates as light does (Source: Merriam Webster)

www.ingramcontent.com/pod-product-compliance
Ingram Content Group UK Ltd.
Pitfield, Milton Keynes, MK11 3LW, UK
UKHW052011041026
15590UKWH00008B/2271

9 781608 882366